3D Printed Science Projects Volume 1

Ideas for Your Classroom, Science Fair, or Home

Second Edition

Joan Horvath
Rich Cameron

Apress®

3D Printed Science Projects Volume 1: Ideas for Your Classroom, Science Fair, or Home, Second Edition

Joan Horvath
Pasadena, CA, USA

Rich Cameron
Whittier, CA, USA

ISBN-13 (pbk): 979-8-8688-0341-3
https://doi.org/10.1007/979-8-8688-0342-0

ISBN-13 (electronic): 979-8-8688-0342-0

Copyright © 2024 by Joan Horvath and Rich Cameron

This work is subject to copyright. All rights are reserved by the Publisher, whether the whole or part of the material is concerned, specifically the rights of translation, reprinting, reuse of illustrations, recitation, broadcasting, reproduction on microfilms or in any other physical way, and transmission or information storage and retrieval, electronic adaptation, computer software, or by similar or dissimilar methodology now known or hereafter developed.

Trademarked names, logos, and images may appear in this book. Rather than use a trademark symbol with every occurrence of a trademarked name, logo, or image we use the names, logos, and images only in an editorial fashion and to the benefit of the trademark owner, with no intention of infringement of the trademark.

The use in this publication of trade names, trademarks, service marks, and similar terms, even if they are not identified as such, is not to be taken as an expression of opinion as to whether or not they are subject to proprietary rights.

While the advice and information in this book are believed to be true and accurate at the date of publication, neither the authors nor the editors nor the publisher can accept any legal responsibility for any errors or omissions that may be made. The publisher makes no warranty, express or implied, with respect to the material contained herein.

    Managing Director, Apress Media LLC: Welmoed Spahr
    Acquisitions Editor: Miriam Haidara
    Development Editor: James Markham
    Project Manager: Jessica Vakili

Cover designed by eStudioCalamar

Distributed to the book trade worldwide by Springer Science+Business Media New York, 1 New York Plaza, New York, NY 10004. Phone 1-800-SPRINGER, fax (201) 348-4505, e-mail orders-ny@springer-sbm.com, or visit www.springeronline.com. Apress Media, LLC is a California LLC and the sole member (owner) is Springer Science + Business Media Finance Inc (SSBM Finance Inc). SSBM Finance Inc is a **Delaware** corporation.

For information on translations, please e-mail booktranslations@springernature.com; for reprint, paperback, or audio rights, please e-mail bookpermissions@springernature.com.

Apress titles may be purchased in bulk for academic, corporate, or promotional use. eBook versions and licenses are also available for most titles. For more information, reference our Print and eBook Bulk Sales web page at http://www.apress.com/bulk-sales.

Any source code or other supplementary material referenced by the author in this book is available to readers on GitHub (https://github.com/whosawhatsis/3DP-Science-Projects). For more detailed information, please visit https://www.apress.com/gp/services/source-code.

If disposing of this product, please recycle the paper

In memory of Zillabell "Jane" Friesen

Table of Contents

About the Authors xiii

Acknowledgments xv

Introduction to the Second Edition xvii

Chapter 1: Math Modeling with 3D Prints 1

   Math Modeling for 3D Printing 2

      OpenSCAD 2

      OpenSCAD Basics 3

      Downloading OpenSCAD 3

      Editing the Models 4

      Math Background 5

      Simple "Blocky" Model 8

      Idiosyncrasies of OpenSCAD 13

   Creating Smoother Surfaces 13

      Limitations and Alternatives 20

      Making a Two-Sided Smooth Surface 20

   Creating Surfaces from an External Data File 22

   3D Printing 23

      Slicing Programs 24

      Printing Considerations 33

   Archives and Repositories 33

   Where to Learn More 35

   Teacher Tips 36

TABLE OF CONTENTS

Science Fair Project Ideas 36
Summary 37

Chapter 2: Light and Other Waves 39
Physics and Math Background 41
    Principle of Superposition 42
    Basic Examples 43
Diffraction and Interference 49
    Light Through One Slit 49
    The Double-Slit Experiment 52
    Diffraction Models 56
Printing Considerations 59
Where to Learn More 61
Teacher Tips 63
Science Fair Project Ideas 64
More Wave Interaction Models 65
Summary 65

Chapter 3: Gravity 67
Universal Gravitation 68
Gravitational Potential Wells 71
    Earth-Moon System Model 71
    Algol Model 79
    Custom Gravity Well Models 81
Orbits 81
    Halley's Comet Orbit Model 85
    Inner Solar System Model 90
    Custom Orbits 94

TABLE OF CONTENTS

Printing Tips 94
Where to Learn More 97
Teacher Tips 97
Science Fair Project Ideas 98
Summary 99

Chapter 4: Airfoils 101
How Wings Work 104
    Flight Forces: Lift, Gravity, Drag, Thrust 104
    Chord, Camber, and Thickness 107
Other Wing Features 109
    Sweep 109
    Taper 109
    Dihedral and Twist 110
    Control Surfaces 110
    Angle of Attack 111
NACA Airfoils 112
    3D Printable Models 113
    The NACA Four-Digit Series 114
    Math of the NACA Model 117
    The Thickness Equation 118
Single Wing OpenSCAD Model 123
Sting and Wings Model 127
    Measuring Lift 138
    Printing Suggestions 143
Classic Airplanes Using NACA Airfoils 144
Where to Learn More 147

TABLE OF CONTENTS

    Visualizing Flow 148
    Building a Student Wind Tunnel 148
    Scaling a Model 149
    Teacher Tips 150
    Science Fair Project Ideas 150
  Summary 151

Chapter 5: Simple Machines 153
  The Machines 155
    Inclined Plane and Wedge 156
    Lever 161
    Screw 169
    Wheel and Axle 175
    Pulley 183
  Printing Suggestions 191
  Where to Learn More 191
  Teacher Tips 192
  Science Fair Project Ideas 192
  Summary 193

Chapter 6: Plants and Their Ecosystems 195
  Botany Background 196
    Water 197
    Sunlight 198
    Nutrients 199
    Plant Communities 199
  The Mathematics of Plant Growth 201
    The Golden Ratio 202
    The Golden Angle 203

TABLE OF CONTENTS

    Fibonacci Sequence 203

    Phyllotaxis 204

The Models.. 204

    Desert Plants 206

    Tropical Jungle Plants 208

    Flowers... 212

Printing the Models 217

    Plant and Flower Models 217

    Jungle Plant Leaf Model 227

    Printing Suggestions 230

Where to Learn More 233

Teacher Tips 234

Science Fair Project Ideas 235

Summary.. 235

Chapter 7: Molecules 237

Chemistry Background.. 238

    The Periodic Table of the Elements 239

    Basic Orbital Shapes 242

    Assembling the Model 250

The Carbon Atom: Hybridized Orbitals 251

    Carbon Atom Model 253

    Assembling the Carbon Atom 256

Water Molecules 258

    The Water Molecule Model 259

    The Carbon Versus Water Molecule Model 262

Crystals 263

    Water Ice 263

    Diamond 269

TABLE OF CONTENTS

Printing Suggestions 271
Where to Learn More 272
Teacher Tips 274
Science Fair Project Ideas 274
Summary 275

Chapter 8: Trusses 277
Engineering Background 278
    Why Triangular Structures? 279
    Forces on Planar ("2D") Truss Members 281
    The Space (3D) Truss 281
    Tensegrity Structures 282
The Models 282
    2D Truss Model 283
    Printing the 2D Truss 289
    Tensegrity Structure Model 289
    Printing the Tensegrity Elements 292
    Assembling the 3-Rod Tensegrity Prism 292
    Hints for Assembling an Icosahedron 298
Where to Learn More 301
Teacher Tips 302
Science Fair Project Ideas 302
Summary 303

Chapter 9: Gears 305
Gears 307
    Gear Ratio 309
    Types of Gears 310
    Sprockets 312

TABLE OF CONTENTS

Gear Models 314
    Gear Set Model 315
    Planetary Gear Model 326
Where to Learn More 327
Teacher Tips 327
Science Fair Project Ideas 328
Summary 328
A Few Last Words About Making Things 329

Appendix: Links 331

Index 337

xi

About the Authors

Joan Horvath and **Rich Cameron** cofounded Nonscriptum LLC in 2015. From their base in Pasadena, California, they consult for educational and scientific users in the areas of 3D printing, maker technologies, and hands-on STEM curriculum development. Joan and Rich are particularly interested in finding ways to use technologies like 3D printing to make math and science easier for everyone, particularly those who learn differently.

Joan and Rich have been prolific authors, with many books from Apress and Make: Community LLC. They have also written for *Make:* magazine and developed online courses for LinkedIn Learning (formerly Lynda.com). Links for all the above are on their website, www.nonscriptum.com.

Joan has taught at the university level in a variety of institutions, both in Southern California and online. Before she and Rich started Nonscriptum, she held a variety of entrepreneurial positions, including Vice President of Business Development at a Kickstarter-funded 3D printer company. Joan started her career with 16 years at the NASA/Caltech Jet Propulsion Laboratory, where she worked in programs including the technology transfer office, the Magellan spacecraft to Venus, and the TOPEX/Poseidon oceanography spacecraft. She holds an undergraduate degree from MIT in aeronautics and astronautics and a master's degree in engineering from UCLA.

ABOUT THE AUTHORS

Rich (known online as "Whosawhatsis") is an experienced open source developer who has been a key member of the RepRap 3D printer development community for many years. His designs include the original spring/lever extruder mechanism used on many 3D printers, the RepRap Wallace, and the Deezmaker Bukito portable 3D printer. By building and modifying several of the early open source 3D printers to wrestle unprecedented performance out of them, he has become an expert at maximizing the print quality of filament-based printers. When he is not busy making every aspect of his own 3D printers better, from slicing software to firmware and hardware, he likes to share that knowledge and experience online so that he can help make everyone else's printers better too.

Acknowledgments

The consumer 3D printing ecosystem would not exist in its current form without the open source 3D printing hardware and software community. We are particularly grateful to Marius Kintel, the main developer and maintainer of OpenSCAD software, and his collaborators for their software, which was used to develop the objects in this book.

The Apress production team made this process as seamless as a complex book like this can be. For the first edition, we dealt most directly with Mark Powers, Michelle Lowman, James Markham, Corbin Collins, and Welmoed Spahr. For the second edition, Jessica Vakili, James Markham, and Miriam Haidara guided the day-to-day aspects of the book's development and production.

We picked a lot of scientists' brains as we thought about how to model some of the concepts in this book. We particularly thank high school teacher Michael Cheverie for his insights into teaching chemistry. Joan's long-suffering astronomer husband, Stephen Unwin, was a huge help as we went back into some basic physics or just tried to get past the 3D modeling equivalent of writer's block. For the first edition, Frank Carsey, Dan Berry, Tim Thibault, and many others helped us out by reading a chapter draft or helping us think through alternative ways that we might model something.

We thank the staff, teachers, and students at the Windward School in Los Angeles and the Institute for Educational Advancement in Pasadena for inspiration and discussions of how students learn. We also were inspired to create these models in part by discussions with people in the community of teachers of the visually impaired, notably Mike Cheverie, Lore Schindler, Yue-Ting Siu, and the participants in the Benetech workshop organized by Lisa Wadors.

ACKNOWLEDGMENTS

As one might expect, our library of 3D prints has evolved over the eight years or so between these two editions. We have drawn on one of the open source math models developed for our *Make: Calculus* book (Make: Community LLC, 2022), which in turn had been somewhat inspired by the math surface model in this book. We note the details in Chapter 1. We also appreciate the many people who had comments and critiques of the models and text in the first edition.

Finally, we are grateful to our families for putting up with our endless brainstorming, kitchen table commandeering, and test runs of explanations both times around.

Introduction to the Second Edition

Almost eight years have gone by since the first edition of *3D Printed Science Projects*. During that time, 3D printers and their software have improved so much that we wound up nearly completely rewriting this book. We are disappointed that the environment in schools has not changed that much. 3D printing largely remains something relegated to key chain fobs.

We hope that this book, in its improved and expanded form, will coax more people into trying to be a bit more ambitious. It is still true that students, parents, and teachers get excited about using a 3D printer, maybe download a model, print it, and then wonder what to do next. Or, they might get into creating models from scratch and get discouraged by the limitations of easier 3D modeling programs or the learning curves of the more capable ones.

We are trying to create a middle path: models that you could just print, but that would be reasonably easy to alter if you wanted to do more. Further, we designed the models so that they would be useful for learning science or math principles while you were changing their features. We wanted to create some seeds of science fair or extra-credit projects—that is, open-ended, meaty explorations that could be explored at a variety of levels.

We were surprised at how hard this turned out to be. Most textbooks and online sites endlessly recycle versions of the same 2D projection of models of science concepts. In each chapter, we have a "Learning Like a Maker" section where we talk about our adventures in defining and implementing the models—which in some cases involved finding online

INTRODUCTION TO THE SECOND EDITION

copies of 1935 manuscripts (signed off by a Wright brother and Charles Lindbergh!). In others, it meant figuring out what to do when scientific experts said that everyone teaches a subject one way, but unlearns it in grad school anyway.

This book presumes that you know a little bit about 3D printing already. If not, Chapter 1 and the resources linked there should get you up to speed. The models are all written using the OpenSCAD free and open source 3D modeling program. If you know how to program in a language like C, Java, or Python, that will help, but it is not strictly necessary to alter the models. Chapter 1 and the OpenSCAD materials linked there will help you out with that too.

We have found that teachers use 3D printers in one of two fundamental ways. Either they want to create a model to pass around in class to help students visualize a concept, or they want students to use a printer either to learn engineering and design per se or to cement physical concepts like levers and gears. Since most of these models would lend themselves to being used either way, we have not included a grade level or explicit lesson plans.

To show our readers who are teachers (in the United States) what we had in mind, though, at the end of most chapters, we suggest Next Generation Science Standards (NGSS) that we think might benefit from these models. These science standards, from the group NGSS Lead States, are documented in *Next Generation Science Standards: For States, By States* (The National Academies Press, 2013). Links are given at the end of relevant chapters. If you are a teacher, you may want to check with your state or school standards as well to see the best fit.

The models span a variety of topics, and we tried to cover as many disciplines as possible. Briefly, here is what you can look forward to:

INTRODUCTION TO THE SECOND EDITION

- Chapter 1 gives you a few options to print many different types of mathematical surface. This ability underlies some of the other models. We also discuss the process of 3D printing in this chapter and how to download the models and review the software the models are written in—the free program OpenSCAD. (The 3D printing material that was in Appendix in the first edition is now integrated into Chapter 1.)

- Chapter 2 creates models of waves to allow you to explore what happens when waves overlap and interfere with each other. You can print yourself a model of Young's famous double-slit experiment to see how light from two slits can interfere.

- Chapter 3 takes us back to Newton and Kepler to learn about planets and stars and how they speed up and slow down in their orbits.

- Chapter 4 allows you to create wings with classic airfoil shapes from the early days of flight. You will be able to make yourself a very simplistic test stand that you can use to measure the lift from the wing with just a fan and a postal scale.

- Chapter 5 lets you create basic models of all the "simple machines"—wedge, inclined plane, lever, pulleys, and screws.

- Chapter 6 allows you to model plants and their ecosystems and to design plants for different environments. Maybe you can create a garden for another planet (or for the Earth after another few hundred years of climate change).

INTRODUCTION TO THE SECOND EDITION

- Chapter 7 lets you begin to explore carbon atoms and how water molecules come together to make two different types of ice crystals.

- Chapter 8 explores 2D and 3D trusses and how you can use them in various explorations.

- Chapter 9 (new in the second edition) builds on Chapter 5's simple machines to explore gears.

- Finally, an appendix aggregates all the links in the book.

We are making the 3D printable models used in this book (although not the book itself!) open source, licensed under a Creative Commons Attribution 4.0 International License (`https://creativecommons.org/licenses/by/4.0/`). That means you can use them for any purpose (including selling prints yourself) and alter and remix them if you credit us. Chapter 1 has some notes about where to find the repositories if you would like to add to these models. We hope these models are the beginning of a set that students everywhere can play with and learn from.

CHAPTER 1

Math Modeling with 3D Prints

Scientific visualization starts with the underlying mathematics. Thus, we are beginning this book on 3D printing for science projects with a chapter on 3D printing mathematical functions. The basic models in this chapter are intended to be a starter set that you alter to 3D print whatever function you like within the boundaries we will get to in a later section.

First, we need to introduce you to the open source program we used to create our 3D printable models, OpenSCAD. Then we use OpenSCAD to create a 3D printed mathematical surface defined by an equation. Then, we cover the 3D printing basics you will need to know to create the models. Finally, we conclude with a brief introduction to 3D printing and resources to help you go farther.

CHAPTER 1   MATH MODELING WITH 3D PRINTS

> **MODELS USED IN THIS CHAPTER**
>
> This chapter uses two different OpenSCAD models. Select 3D printable STL example files are included in the repository as well, which we explain in the chapter text. The OpenSCAD models are
>
> - `BlockyMath.scad`: This model creates a blocky 3D surface, shown in Listing 1-1.
>
> - `surfaceprint.scad`: This model creates a smooth 3D mathematical surface based on an equation of the form $z = f(x, y)$. It is shown in Listing 1-2.
>
> You also need to install the OpenSCAD program, as we describe in the chapter. As of this writing, OpenSCAD does not work on a Chromebook or similar tablet. A MacOS, Windows, or Linux computer is required.

Math Modeling for 3D Printing

It seems like it should be easy to just put an equation into a program and have a 3D printer "draw" it, like some sort of pen plotter. However, if a 3D printer head just tried to follow an equation, it would have no way of knowing how to avoid material that had already been laid down, so we go about it in a bit more roundabout way.

OpenSCAD

3D printers require a several-step process from the first idea to a finished print. First, you need to develop a 3D model. Models in this book are built using OpenSCAD (www.openscad.org), a free, open source 3D solid modeling program. Then, other software takes this file and slices it into

CHAPTER 1   MATH MODELING WITH 3D PRINTS

layers, which the printer will then create one at a time, typically from the bottom up. This software is called a "slicer," and we will discuss common features of these programs in the "3D Printing" section of this chapter. The output of the slicer is what the 3D printer needs to make something.

Note   Models in this book were optimized for printers that use filament and were designed to work well even on a printer that might not be tuned perfectly. If you have a printer that uses liquid resin or powder, you might have to make some adjustments.

OpenSCAD Basics

OpenSCAD allows you to encode geometrical models in a programming language that is like those in the C/Java/Python family. Good documentation is available by clicking the Documentation button on the OpenSCAD site's home page. We will denote OpenSCAD model code by showing it in code font. We want to acknowledge and thank Marius Kintel and the many other contributors to and maintainers of the program.

Downloading OpenSCAD

You can download OpenSCAD from www.openscad.org. Install the program per the instructions on the download site. OpenSCAD is available in versions for MacOS, Windows, and Linux. The models in this book were tested with version 2021.01 for MacOS. If you are a longtime OpenSCAD user and have an older version than that, you may need to update to the current version to be able to run the models in this book, which take advantage of some recently added features.

If you scroll down further on the downloads page, you can find OpenSCAD's nightly builds. These include some optimizations that dramatically increase rendering speed on some of our models, but they are not fully tested, and depending on when you download them, there may be significant software bugs. OpenSCAD has an excellent user manual at www.openscad.org/documentation.html. Note that there are block-coding-based versions of OpenSCAD, but as of this writing, they do not support all the functionality needed for our models.

Editing the Models

Although the models in this book can be printed with their default values, the intent is that you use them as a starting point and make them your own. To edit one of the models in this book, first you would obtain the relevant OpenSCAD (.scad) file for the model you are interested in. See the "Archives and Repositories" section of this chapter for how that works.

Once you have a file you want to use, open OpenSCAD, click File ➤ Open, and open the .scad file. You may see four panes: one with code in it (the Editor), two labeled Customizer and Console that we will get to later, and one that starts off with just coordinate axes, which will in due course display a preview of our math model. (You might have to explicitly open one or more of these windows depending on your settings.)

Each of these panes can be resized, moved around, or hidden within the window, so your window may look a little different than the ones in our screenshots. If there are any of those panes that you do not see, go to the View menu, and uncheck the relevant Hide line (e.g., uncheck Hide Editor if you do not see the Editor window).

Figure 1-1 shows a very simple model that builds a surface out of blocks, which we explore in the next few sections. By default, OpenSCAD also previews the model for you. This can be turned off by unselecting Design ➤ Automatic Reload and Preview if you are working with models that take a while to render and you want to control it.

CHAPTER 1   MATH MODELING WITH 3D PRINTS

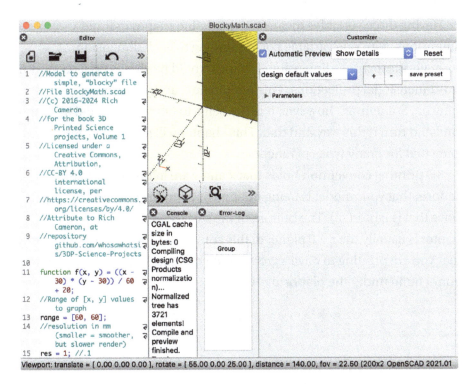

Figure 1-1. Opening a model in OpenSCAD

Math Background

We have tried to avoid too many equations in this book. We do assume that you understand basic algebra, Cartesian coordinates, what it means to raise a number to a power or take a root of it, and trigonometric functions like sine, cosine, and tangent. But, with that said, science often requires some math to describe it, and that math then is used to create models.

We also presume you know what a mathematical function is—a relationship among a number of variables. In this case, we are dealing with functions using three variables, which we will call x, y, and z. Function notation looks like this: $z = f(x, y)$. All that means is that our variable, z, can be computed for any given pair of values for the x and y variables.

5

CHAPTER 1   MATH MODELING WITH 3D PRINTS

Having three variables means we can define shapes in three dimensions, with one variable corresponding to each dimension. Normally these three-dimensional shapes would be shown on a page with two-dimensional projections. This is often fine, and you can see what is going on. Sometimes, however, it really helps to hold a 3D model in your hand and turn it this way and that. This chapter will get you started on doing that for many types of functions.

3D printing convention holds that x and y are in the plane of the platform that your model is being built up on, and z is the vertical height above that (Figure 1-2). The bottom of the surfaces generated in this chapter is usually the $z = 0$ plane. In this convention, you always transform what you are printing to have z greater than or equal to zero since you cannot build under the platform.

CHAPTER 1   MATH MODELING WITH 3D PRINTS

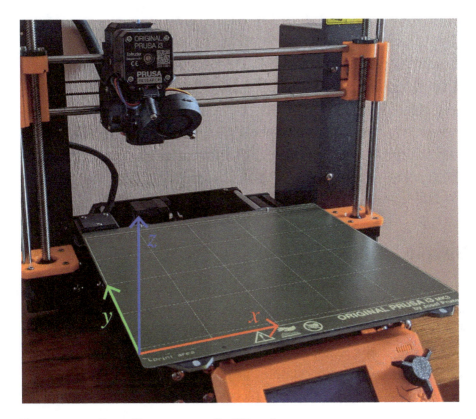

Figure 1-2. *Coordinate axes of a 3D printer*

In other words, if you know that z would be negative for some values of x and y that you want to use, you may have to add an offset to your equation so that z is always greater than zero and remember that the offset is there when you think about what your model represents. In the next section, we get you started with a model entirely in OpenSCAD that creates surfaces of functions

```
z = f(x,y),
```

where x and y are the plane of the 3D printer's build platform and z is the height of the surface above that plane.

CHAPTER 1   MATH MODELING WITH 3D PRINTS

Tip   The maintainer of OpenSCAD has co-authored a book on learning to code with OpenSCAD, *Programming with OpenSCAD: A Beginner's Guide to Coding 3D Printable Objects* by Justin Gohde and Marius Kintel (2021: No Starch Press). You might find that a good resource for more depth on OpenSCAD's ins and outs. We also have a more in-depth review of OpenSCAD as a tool for learning math in our book *Make: Geometry* (2021: Make: Community LLC).

Simple "Blocky" Model

The simplest way to print a 3D surface is to compute the height of the surface on a regular grid of points. Then, we create a rectangular solid that is a small square at the base, with the height equal to the value of the function at the center of each of these small squares. The model in Listing 1-1 is an extremely simple one that will create an STL file with a surface of small rectangular solid pieces. The function in this example is

$z = f(x, y) = 0.01 (x - 50) (y - 50) + 30$,

and the 3D print will go from $x = 0$ to $x = 60$ and $y = 0$ to $y = 60$. This creates a "saddle point" structure, as shown in Figure 1-6. The x and y dimensions of the model are determined by the range = [x, y] variable in millimeters, with z height computed, also in millimeters. If the resulting structure is too big (by default, 60 mm by 60 mm on the bottom, or a bit over 2 inches square), then you can scale the whole piece in your 3D printing software.

The values of x and y both start at 0 and go to the values specified by the range vector (read as range[0] for x and range[1] for y). The res variable, also in millimeters, lets you control the spacing of the calculated points, allowing you to make a smoother surface at the cost of additional processing time.

CHAPTER 1   MATH MODELING WITH 3D PRINTS

Table 1-1 lists the model parameters that can be changed. Figure 1-3 shows a surface generated this way. Notice a pattern on the top surface; this is a printing artifact that we will talk about in the "3D Printing" section.

Table 1-1. Blocky Surface Model Variables

Variable	Default Value and Units	Meaning
f(x,y)	mm	Function to graph.
range	[60, 60] mm	Maximum x/y values to graph. Determines the size of the print in those dimensions.
res	1 mm	Size of the boxes in the graph. Making this smaller produces a smoother surface but takes longer to process.

Figure 1-3. *The blocky surface, as printed*

9

CHAPTER 1   MATH MODELING WITH 3D PRINTS

Listing 1-1. OpenSCAD Model to Generate a "Blocky" Surface (file `blocky.scad`)

```
//Model to generate a simple, "blocky" file
//File BlockyMath.scad
//(c) 2016-2024 Rich Cameron
//for the book 3D Printed Science projects, Volume 1
//Licensed under a Creative Commons, Attribution,
//CC-BY 4.0 international license, per
//https://creativecommons.org/licenses/by/4.0/
//Attribute to Rich Cameron, at
//repository github.com/whosawhatsis/3DP-Science-Projects

function f(x, y) = ((x - 30) * (y - 30)) / 60 + 20;
//Range of [x, y] values to graph
range = [60, 60];
//resolution in mm (smaller = smoother, but slower render)
res = 1; //.1

for(x = [0:res:range[0]], y = [0:res:range[1]])
  translate([x, y, 0])
    cube([res + .001, res + .001, f(x, y)]);
```

To see this surface in OpenSCAD, use the menu item Design ▶ Preview. You can also do this by clicking on the little box outlined with dashed lines with a double arrow on it. Make any changes you feel you need to make by editing the text in the Editor and click Preview again to see if you have created what you intended. Repeat until you think you are done.

Note   You cannot make any changes other than by changing the model code in the Editor. There is no drag and drop capability in OpenSCAD.

In OpenSCAD, `Preview` creates an object you can view but cannot export. It is a lot faster than a full render, which can take a long time for some of the models in this book. Use this to preview models as you are making changes. When you have your final model, go to `Design` ➤ `Render` (or click the button with the solid-outlined cube next to the preview button) to create a model that can be exported for 3D printing. This is called an *STL file* (more on this in the "3D Printing" section). You can export an STL file by clicking `File` ➤ `Export` ➤ `Export as STL`. The editor pane also has an STL export button that you can use.

There is another way to make changes to some of the models in this book. This is by using the `Customizer`. In Figure 1-1, you can see the `Customizer` pane has a pull-down menu labeled `Parameters`. If we expand that pull-down menu (Figure 1-4), we see the `range` and `res` boxes. We can change those values with the pull-down menu or type a new one. These values will then supersede the ones in the code until you close and reopen the file.

Many of the models in later chapters are designed so that you only need to alter parameters for the model in the `Customizer`, which avoids needing to touch the code that makes up the OpenSCAD file. In other cases, we talk you through how to make those changes. Note that the `Customizer` does not support mathematical expressions, so you still need to use the code editor for changing the function.

CHAPTER 1   MATH MODELING WITH 3D PRINTS

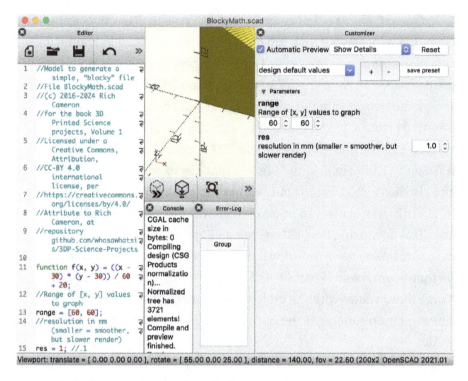

Figure 1-4. Parameters expanded

The Console window will give you information about generating the preview or final versions of the model. This includes any errors encountered while processing the model. Some models also produce additional information (using the echo() module, which is like print() or printf() in other languages) that will be displayed here.

Finally, if you look at File ➤ Examples in OpenSCAD, you will see many sample models. Open a few and play with them to get comfortable before moving on to the models in this book. Other people creating math or science models (see links at "Where to Learn More" later in this chapter) have used more sophisticated mathematics modeling programs, but our desire here is to make models in our books completely accessible and free so that you can get started without investing in software, at least at the beginning of your explorations.

CHAPTER 1   MATH MODELING WITH 3D PRINTS

Idiosyncrasies of OpenSCAD

If you are a programmer, OpenSCAD can be a little disconcerting because its syntax looks like that of the C/Java/Python family of languages. But it is not a full programming language and has a few idiosyncrasies. The biggest one is that OpenSCAD does not have true variables, as one would define them in other programming languages. The variables in our models are best thought of as constants. You can assign another value to a variable, but (as would be true in algebra)

```
y = y + 1;
```

is not a valid statement in OpenSCAD. See the manual section on variables for details and examples at https://en.wikibooks.org/wiki/OpenSCAD_User_Manual/General#Variables.

Functions in OpenSCAD are functions in the mathematical sense. They return a value but cannot perform other tasks beyond a single mathematical formula along the way. OpenSCAD has "modules" that are closer to what most experienced programmers will expect from a function.

Finally, trigonometric functions (like sin() and cos()) expect their arguments to be in degrees, not radians. This is unusual among modern programming languages, which typically expect arguments in radians. Be careful to adapt any preexisting algorithms to allow for that, remembering that 2π radians equals 360 degrees.

Creating Smoother Surfaces

Now, we will create a surface with a flat base that is better than our blocky approximation. This model was written to be simple and easy to alter, which means that it does not check for complicated problems, like functions that go to infinity or other mathematically bad behavior. It does, however, let you input a function f(x, y) as you can see in Listing 1-2, with parameters you can change listed in Table 1-2.

CHAPTER 1   MATH MODELING WITH 3D PRINTS

It uses OpenSCAD's `polyhedron()` module to create a flat-bottomed "slice" of a surface, like a chunk cut out of a mountain range. The function `f(x, y)` and the `range` variable work just like in the preceding `blocky` model. The `res` variable also works similarly by controlling the approximate spacing of calculated points in a triangular grid. This triangular grid creates a smoother surface with fewer artifacts than the square grid used in the first edition of this book. This model adds the `blockymode` variable, which makes it function like the simpler `blocky` model when set to `true`. The `t` variable lets you create a two-sided surface by setting it to a nonzero value, representing the thickness separating the two surfaces. Setting `t = 0` produces a flat base at `z = 0`.

We show this surface in Figure 1-5. Note that the surface has a regular pattern called moiré fringes. These artifacts can appear on shallow horizontal surfaces. See the section on Reorienting Models for Better Quality in this chapter for how to minimize this effect.

You might have to scale your surface, depending on the natural units for your x and y variables. For example, if you wanted to graph values of one full period of sine or cosine (again noting that unlike most programming languages, OpenSCAD uses degrees instead of radians for these functions), you could set `range = [360, 360]`. This would result in exporting a model with a 360 mm square base, which would probably need to be scaled down in your printing software.

Alternatively, if you wanted the model to be scaled to a 100 mm square, you could instead set that size and scale the variables in your function, using an expression like `sin(x * 100 / 360)`. This method would process faster than using `range = [360, 360]` and then scaling down while using the same res value, because the larger range means that there is more computing to do.

CHAPTER 1   MATH MODELING WITH 3D PRINTS

Figure 1-5. *Saddle function*

Table 1-2. *Smooth Surface Model Variables*

Variable	Default Value and Units	Meaning
f(x,y)	mm	Function we are plotting.
t	0 mm	Thickness of the surface in the z direction. Use 0 for a flat base at z = 0.
range	[60, 60] mm	Maximum x/y values to graph. Determines the size of the print in those dimensions.
res	0.25 mm	Approximate size of the triangles in the graph. Making this smaller produces a smoother surface but takes longer to process.
blockymode	false	When true, works like `blocky.scad`.

CHAPTER 1 MATH MODELING WITH 3D PRINTS

If you want to graph over a range that does not end at zero, you can also adjust this inside your function. For example, if you wanted a 100 mm square model showing values of x and y from -1 to 1, you could set range = [100, 100] and use

(x / 50 - 1) and

(y / 50 - 1)

in your equation. With res = 1, this would mean that your variables increment by 0.02 per facet.

Scaling the function's z axis can be accomplished by multiplying your whole function (usually in parentheses) by a scaling factor or by scaling the model in your slicing software. Likewise, you can change the height of z = 0 (to show negative values, for example) by adding a constant. Our example function demonstrates both of these.

Caution Because we are creating a flat bottom, the equation being represented here is actually 0 <= z <= f(x, y). As a result, z = f(x, y) must be greater than zero everywhere for the flat-bottomed (t = 0) version. A model will still be produced if there are z values that are less than zero, but it will be an invalid model. Even if your slicer accepts the model, it will not print easily. Some slicers might just not work at all.

Listing 1-2. The Basic OpenSCAD Model to Create a 3D Print of a Surface (file surfaceprint.scad)

```
//OpenSCAD model to print out an arbitrary surface, z = f(x,y)
//Either prints the surface two-sided with t = thickness
//Or if t = 0, prints a top surface with a flat bottom
//File surfaceprint.scad
```

CHAPTER 1   MATH MODELING WITH 3D PRINTS

```
//(c) 2016-2024 Rich Cameron
//for the book 3D Printed Science projects, Volume 1
//Based on triangleMeshSurface.scad
//from github.com/whosawhatsis/Calculus
//Licensed under a Creative Commons, Attribution,
//CC-BY 4.0 international license, per
//https://creativecommons.org/licenses/by/4.0/
//Attribute to Rich Cameron, at
//repository github.com/whosawhatsis/3DP-Science-Projects

//Thickness along z axis. t = 0 gives a flat base at z = 0
t = 0;
//Range of [x, y] values to graph
range = [60, 60];
//resolution in mm (smaller = smoother, but slower render)
res = .25;
blockymode = false;

function f(x, y) = ((x - 30) * (y - 30)) / 60 + 20;

s = [
  round((range[0] - res/2) / res),
  round(range[1] / res * 2 / sqrt(3))
];
seg = [range[0] / (s[0] - .5), range[1] / s[1]];

function r(x, y, cx = range[0]/2, cy = range[1]/2) =
  sqrt(pow(cx - x, 2) + pow(cy - y, 2));
function theta(x, y, cx = range[0]/2, cy = range[1]/2) =
  atan2((cy - y), (cx - x));
function zeronan(n) = (n == n) ? n : 0;
```

17

CHAPTER 1   MATH MODELING WITH 3D PRINTS

```
points = concat(
  [for(y = [0:s[1]], x = [0:s[0]]) [
    seg[0] * min(max(x - (y % 2) * .5, 0), s[0] - .5),
    seg[1] * y,
    zeronan(
      f(
        seg[0] * min(max(x - (y % 2) * .5, 0), s[0] - .5),
        seg[1] * y
      )
    )
  ]], [for(y = [0:s[1]], x = [0:s[0]]) [
    seg[0] * min(max(x - (y % 2) * .5, 0), s[0] - .5),
    seg[1] * y,
    t ? zeronan(
      f(
        seg[0] * min(max(x - (y % 2) * .5, 0), s[0] - .5),
        seg[1] * y
      )
    ) - t : 0
  ]]
);
*for(i = points) translate(i) cube(.1, center = true);
function order(point, reverse) = [
  for(i = [0:2]) point[reverse ? 2 - i : i]
];
function mirror(points, offset) = [
  for(i = [0, 1], point = points)
    order(
      point + (i ? [0, 0, 0] : [offset, offset, offset]),
      i
    )
];
```

18

```
polys = concat(
  mirror(concat([
    for(x = [0:s[0] - 1], y = [0:s[1] - 1]) [
      x + (s[0] + 1) * y,
      x + 1 + (s[0] + 1) * y,
      x + 1 - (y % 2) + (s[0] + 1) * (y + 1)
    ]
  ], [
    for(x = [0:s[0] - 1], y = [0:s[1] - 1]) [
      x + (y % 2) + (s[0] + 1) * y,
      x + 1 + (s[0] + 1) * (y + 1),
      x + (s[0] + 1) * (y + 1)
    ]
  ]), len(points) / 2),
  mirror([for(x = [0:s[0] - 1], i = [0, 1]) order([
    x + (i ? 0 : 1 + len(points) / 2),
    x + 1,
    x + len(points) / 2
  ], i)], len(points) / 2 - s[0] - 1),
  mirror([for(y = [0:s[1] - 1], i = [0, 1]) order([
    y * (s[0] + 1) + (i ? 0 : (s[0] + 1) + len(points) / 2),
    y * (s[0] + 1) + (s[0] + 1),
    y * (s[0] + 1) + len(points) / 2
  ], 1 - i)], s[0])
);

if(blockymode)
  for(x = [0:res:range[0]], y = [0:res:range[1]])
    translate([x, y, 0]) cube([res, res, f(x, y)]);
else polyhedron(points, polys, convexity = 5);
```

Limitations and Alternatives

To keep the model simple, transparent, and easy to understand, we have included only minimal error checking for special cases. There is code to detect division by zero and insert a zero value instead. If you have a function that otherwise goes to infinity or has some sort of discontinuity, you may need to come up with some fix. You could, for example, create a branch in the definition of `f(x, y)` to handle cases that are poorly behaved in the mathematical sense. You may want to, for example, enclose your function in `max(0, ...)` to prevent it from creating negative values that would make your file unprintable.

We could only test a few cases, and there are an infinite number of options. As with any print, you should check the software model of your print before committing it to be sure that it worked for your particular example. As discussed earlier, z has to be greater than zero.

Another parameter in the model is `blocky`, which in this example is set to false. Setting `blocky = true` will create what we got in our first example created by `blocky.scad`.

OpenSCAD's math functions will probably look familiar if you are a programmer, but some of them will not be what you are expecting if you do not have that experience already. Exponents, for example, take the form (base ^ exponent) or pow(base, exponent) rather than using a superscript, and a square root uses the `sqrt()` function instead of a radical sign. You can find a (nearly) complete listing of the mathematical functions available in OpenSCAD in its documentation, https://en.wikibooks.org/wiki/OpenSCAD_User_Manual/Mathematical_Functions.

Making a Two-Sided Smooth Surface

Although it is convenient to be able to visualize the top of a surface in 3D, sometimes it is even better to see it from both sides. However, you might think that then you will have to put a lot of support under the surface and wind up more or less with the same thing. The way out of this is to print the surface *sideways*. As long as the surface does not contain slopes that are too steep, you can use the following process to print a surface that is a few millimeters thick (2 mm in this example, the minimum we recommend) and print it on its side.

Run the same model in OpenSCAD as in the previous section (after changing the t parameter to equal the desired thickness in mm), scale it if you like (be sure the thickness does not go below about 2 mm), and rotate it 90 degrees about either the x or y axis. Look at your model to determine which rotation will have the fewest overhangs. If the excursions in f(x, y) were big, you might have to use some support this way, but probably far less than if you laid the surface flat. As a bonus, z can be negative in this version.

The only difference between this case and the previous section is that the parameter t was a nonzero positive number. In this case, we set it to 2. Thus, we generated a file that had the dimensions 60 by 60 mm in x and y and was 2 mm thick in z. Figure 1-6 shows the finished print still on the printer, held down in part by its attached *brim* (see the "3D Printing" section of this chapter).

Caution If you scale the surface, you have to be sure that the piece remains at least 2 mm or so thick after scaling. If you want something about 50 mm across, for example, and will be scaling it down by a factor of 2, you should start with t = 4. Otherwise, the surface might get too thin to print in places, and it will develop holes that might make the print fail.

CHAPTER 1   MATH MODELING WITH 3D PRINTS

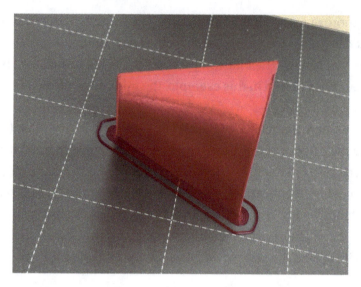

Figure 1-6. *The saddle surface print as it finished*

Creating Surfaces from an External Data File

The processes we have used in the chapter so far have the virtue that everything was controlled entirely within OpenSCAD. However, since OpenSCAD does not have variables in the normal coding sense of the word, it is hard to do things like scale the surface dynamically or do any of the other things you might want to do for a complex mathematical model. (As we noted earlier in the section on OpenSCAD idiosyncrasies, it has variables as a mathematician might think of them. Most programming languages would call them constants.) You might have another piece of software creating your surface data points, and all you want to do is essentially plot them out as a 3D surface, either with a flat base or as a thin surface.

OpenSCAD's surface( ) module can plot a surface from a file of data points. You could use a spreadsheet or for that matter a text editor to create your file of values. We used Microsoft Excel to create a file of

CHAPTER 1 MATH MODELING WITH 3D PRINTS

tab-delimited numbers to plot, saved in a file called `file.txt`, with each line of tab-delimited values corresponding to a fixed y value as x varies. The `surface()` module will also accept an image file in PNG format, using the grayscale luminance of each pixel for the height. The entirety of the OpenSCAD code to do this is:

```
surface(file = "file.txt");
```

If any of the values in the data file are negative, the `surface()` module will treat the most negative one as the value for the bottom of the base and offset other positive values accordingly. It assumes that the data points are in a 1 mm square grid.

3D Printing

Now that we know how to create (or at least open) a model in OpenSCAD, let's see how we go from there to a 3D printed object. Computer-aided design (CAD) programs like OpenSCAD can create several different file types, intended for different purposes. OpenSCAD can only edit its own input files (`.scad` files). The format output for 3D printing is called an *STL file*, which stands for either *standard tessellation language* or *stereolithography* (the first 3D printing technique), depending on who you ask.

Tessellation is the process of completely covering a surface with one or more shapes. STL files are made up of many triangles that are used to create a mesh over the surface that was created by the CAD program. (The smooth-surface model in this chapter is doing something similar.) The STL file is made up of the coordinates of the vertices of these triangles, plus their orientation in space. The inside of the object is not modeled. An STL file also does not preserve any information about color or material properties desired. It is, despite those shortcomings, the commonest input into the next step in the 3D printing process: a slicer.

23

CHAPTER 1   MATH MODELING WITH 3D PRINTS

Slicing Programs

Once you have exported your STL file from OpenSCAD, you need to run a program that can convert the model into commands to drive your 3D printer. We used the open source slicer PrusaSlicer (at www.prusa3d.com, click Software). It is freely available and supports many printers, although it was designed for printers by Prusa Research. There are other multi-printer systems out there too, both free and with a subscription.

Many printers have proprietary slicing software optimized for them. Sometimes these are based on an open source slicer, sometimes not. If you are new to 3D printing, we suggest you start with the slicer recommended for your printer with the default settings. Slicers usually have an intimidating number of options, but there are just a few settings you might want to change when you print models from this book.

Because slicers tend to be updated frequently, and there are so many now, we are not going to show you step-by-step the process of using a particular one. Most have good documentation, and our book *Mastering 3D Printing: 2nd Edition* (Apress, 2021) is an in-depth reference if you want to learn more. We have designed the models in this book to print as easily as possible. However, you might want to tweak a few of the settings for some of the prints, as follows.

Materials

The models for the book as photographed were for the most part printed in PLA (polylactic acid), a material made from corn or sugarcane. It is the easiest material to work with and is nearly universally supported on lower-end 3D printers. It does have the disadvantage that prints made with it will warp if they get hot, for example, on a car dashboard in summer.

A few prints were made in PETG (polyethylene terephthalate glycol), which is similar to the PET plastic used in water bottles and the clear plastic packaging of some groceries. PETG is more heat resistant, and prints made with it can be more translucent. Slicers need to be told which material the printer will use because different temperatures and other settings will be needed. If you are just beginning to use a 3D printer, we suggest you stick to PLA.

Layer Height

Layer height is the thickness in the vertical direction of each layer of your print, typically in millimeters. We usually go for 0.2 mm, or maybe 0.1 or 0.15 for finer detail. The smaller the layer, the better the surface quality, but the longer it takes to print. Your printer probably has a stated minimum layer height, and it is best to stick to that.

Using Support and Infill

Since a filament-based 3D printer builds up prints from a platform, if a piece sticks out sideways higher up on a print, that part will just fall down if *support material* is not built ahead of time. Most of the models in this book have been designed to avoid support. Some of the botanical models in Chapter 6 might require it, depending on your settings. You can look in the previewer for your slicer to see if it looks like the overhangs will be too large to manage without support.

There is a rule of thumb that a slope can overhang by about 45 degrees before support is necessary. However, sometimes you can push your luck. Figures 1-7 through 1-9 are of one of the flower models in Chapter 6. It was printed without support. You can see that it overhangs significantly, but we suspected the organic shape would survive without support. You can see a little raggedness on the farthest-out petals, particularly in the view from below. On the bottom, plastic would have been deposited partially in the air without support, which is why it is particularly rough there.

CHAPTER 1 MATH MODELING WITH 3D PRINTS

Infill is a structure that the printer automatically creates on the inside of your print. An STL file is just triangles modeling the shell of your structure. Infill acts like support inside the print. 3D prints are rarely solid. We typically use from 12% to 20% infill in our practice. You can pick different infill patterns, too, which might vary a little in strength and (if the print is translucent) esthetics.

Figure 1-7. *The flower from Chapter 6 from above*

CHAPTER 1   MATH MODELING WITH 3D PRINTS

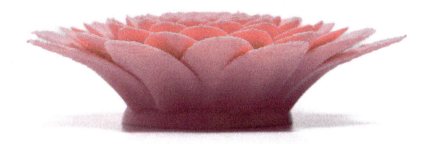

Figure 1-8. *The flower from Chapter 6 from the side to show the extent of unsupported overhangs*

Figure 1-9. *The flower from Chapter 6 from below*

CHAPTER 1   MATH MODELING WITH 3D PRINTS

In general, the less support you generate (and then pick off), the better. If you have something complex, you may have to create support everywhere. This creates support as needed, including in nooks and crannies of the model where it may be hard to remove. Be sure to preview your print first to see how it looks.

Figure 1-10 shows a different flower model that needed support since it had more extreme overhangs. We used the "organic" and "paint on" supports in PrusaSlicer to create the supports here. Most slicers will allow you to put support only where needed or to selectively remove it. That allows you to avoid having support places you do not want.

Figure 1-10. *The completed flower-with-support print, with the support still on it*

Making a Print Stick

Some of the models in this book depend on your printer's bed being smooth and flat so that multiple pieces can be printed and then arranged flush with each other (like the airplane wing with the sting in Chapter 4 or the water molecule halves in Chapter 7).

If your printer's bed is not very flat, you can print these parts on a raft. A raft is a thin layer that prints first on the platform, and then the model prints on top of it. If things are not fitting together well, adding a raft is an option. You will need to remove the raft, though, which may be difficult to do cleanly if your raft settings have not been tuned to allow it to release from the print. You may want to test it with a smaller piece to make sure the raft will peel away in one piece before printing a larger model on a raft.

Note that many printer manufacturers recommend using a glue stick or tape on the printer's platform. This is not to make the print stick. Rather, these act as a release layer so that the print does not stick too well. Many modern printers have a removable, flexible platform. If so, removing it and flexing it will usually pop off the print. If a print is really on there, devices like small spatulas can be used to gingerly get one edge up and work it off from there. Be careful not to damage your print bed since any dings will be imprinted into all future prints!

Skirt and Brim

Your slicer may offer you the option of a skirt or brim. A skirt is a line drawn around the model's first layer to prime the nozzle. If you make a skirt attached (sometimes by selecting to place it 0 mm away from the model), it is usually called a brim (as in hat brim). A brim can help hold prints onto the print bed, though it will not help flatten the bottom of a print the way a raft will.

Reorienting Models for Better Quality

Sometimes a print will work better if it is oriented differently than it is created. You might be able to avoid support that way. Layer lines are usually thinner than the line of plastic the printer is extruding. This means that you might want details on the side of a print, rather than on top. For example, Figure 1-11 compares the print from Figure 1-3 as printed and then as rotated 90 degrees.

CHAPTER 1   MATH MODELING WITH 3D PRINTS

Figure 1-11. *Two blocky surface prints compared*

Figure 1-12 does the same for the smoother surface in Figure 1-5. The surface printed on its side (in the center) is clearly much cleaner and does not have the interference patterns that we otherwise get with small differences interacting with layer lines. We also have printed it on its side and made it hollow—a *vase print* (on the right of Figure 1-12). This can be a quick and attractive way of getting a surface print, which we describe in the next section.

CHAPTER 1   MATH MODELING WITH 3D PRINTS

Figure 1-12. *Three smooth surface prints compared (L: printed with surface up, center: printed surface on the side, R: surface on the side and hollow)*

Vase Mode

One final option to get a good surface quickly is to print the model from surfaceprint.scad (rightmost in Figure 1-12, close-up in Figure 1-13) on its side, but hollow and with an open top, like a vase. PrusaSlicer calls this setting spiral vase. Others might call it something like "spiralize outer contour" or "vase mode." In any case, since there is so little plastic, it can be a quick way to model a surface. You can also accomplish the same thing by telling your slicer to use zero infill and no top layers, sometimes called "old-school vase mode" by grizzled 3D printing veterans.

CHAPTER 1   MATH MODELING WITH 3D PRINTS

Figure 1-13. Close-up of vase printed surface

Cautions   Even with a consumer-level 3D printer, there are always some cautions to keep in mind. Always have adequate ventilation when running your 3D printer and keep children and inquisitive pets away from the hot nozzle.

Note that 3D prints are not food-safe unless you are using a printer specifically designed for it or do some sophisticated post-processing. There are two reasons for this: first, that the nozzle may leave residues on the print. Second, the layer lines can trap bacteria if a print is used repeatedly for food.

Finally, these models are educational aids and not toys. Some have small parts and are not suitable to have around young children.

CHAPTER 1   MATH MODELING WITH 3D PRINTS

Printing Considerations

Since the models in these chapters were designed to have a flat bottom, they 3D print easily. There will not be any overhangs. If your function would get very tall, you may want to scale it down by multiplying the equation by a constant. To check that the function is not getting too big or going negative in the z direction, you can either just run it in OpenSCAD and look at the result or graph it conventionally yourself to see what it looks like. (The model does not check it for you.)

For the most part, the models in this book were printed in PLA with a layer height of 0.2 mm with 15% infill. Your printer defaults may be different than this, and you may want to do a few tests with these simpler prints in this chapter to establish your favorite basic settings.

Archives and Repositories

There is a link for the source for the OpenSCAD models on this book's copyright page. The files archived there are the ones shown in the book. To download the models from GitHub, go to the repository site, github.com/whosawhatsis/3DP-Science-Projects. On the right-hand side is a button that says "Release" and a button next to it that says "Latest." Click that and you will go to a page called Assets. Download the 3DP-Science-Projects.zip file which contains all the .scad files noted in the book, and select STL files to print.

The models are licensed with a Creative Commons CC-BY 4.0 International License, as described at creativecommons.org/licenses/by/4.0/. This means that you can use them freely, but you need to give credit to (attribute) them. Attribution should be as follows:

33

CHAPTER 1   MATH MODELING WITH 3D PRINTS

"From the repository of the book *3D Printed Science Projects Volume 1, 2nd Edition* by Joan Horvath and Rich Cameron Located at `https://github.com/whosawhatsis/3DP-Science-Projects`, license CC-BY as described at `https://creativecommons.org/licenses/by/4.0/`."

If you develop any new models around these, we hope you will add them to the open source repositories and help build out a community of scientific learners. In that spirit, note that the version of `surfaceprint.scad` in the second edition of this book has heritage to the model `triangleSurfaceMesh.scad` from our book *Make: Calculus*, repository github.com/whosawhatsis/Calculus.

THINKING ABOUT THESE MODELS: LEARNING LIKE A MAKER

In each chapter of this book, we talk about what we learned while we were developing the models so that you can think about what you would like to do with them. This chapter is a little different from the rest in that the models here are an underlying tool that is useful for many of the other chapters of the book. In Chapter 2, for example, we will explore waves and fields and will draw on the ability to plot out 3D functions that are usually shown as 2D projections.

In our case, we had to think hard about what types of functions would most usefully be visualized this way and how to balance an overly complex model that would be hard for you to alter versus a reasonably capable one. We changed the more complex model we had in the first edition of this book to the more straightforward ones here for this edition.

We found, as we worked on some of the models later in this book, that often everyone uses the same 2D projection of a 3D model and that creating the entire model literally gives you a different perspective.

CHAPTER 1   MATH MODELING WITH 3D PRINTS

Where to Learn More

The math surfaces in this chapter underlie many of the science models in the upcoming chapters. If you or your students need a little math refresher, Joan's personal favorite place to send students to learn more about math (but not tied to 3D printing) is the Khan Academy, at www.khanacademy.org.

To learn more about 3D printing math models, you can check out our series of books *Make: Geometry* (2021), *Make: Calculus* (2022), and *Make: Trigonometry* (2023), all published by Make: Community LLC. The *Make: Geometry* book has a section on using OpenSCAD as a geometry simulator to learn both the program and math, as well as a long discussion of how to use OpenSCAD more generally.

This book presumes you are generally familiar with 3D printing practices. If not, you can learn how to use a printer from our previous book, *Mastering 3D Printing, 2nd Edition* (Apress, 2021). The majority of the prints in this book were created with polylactic acid (PLA) plastic on an Original Prusa i3 Mark 3S+ printer, using PrusaSlicer for the slicing program.

Finally, if you want to see some other 3D printed mathematics, have a look at the sites of Elizabeth Denne (of Washington and Lee University), which mostly use the program Mathematica for modeling: mathvis.academic.wlu.edu.

Sculptors Bathsheba Grossman and Henry Segerman create mathematics-based sculptures, available at their respective sites: www.bathsheba.com and www.shapeways.com/shops/henryseg. Paul Nylander (http://bugman123.com) has a variety of math- and science-oriented models on his site. And of course, you can always search on the various websites of 3D models (such as www.thingiverse.com and www.printables.com) for even more math models.

35

CHAPTER 1   MATH MODELING WITH 3D PRINTS

Teacher Tips

In later chapters, we give some suggestions about alignments to science standards. Here, rather than give an enormous list of possible links, we will just note that pretty much any level of math that can benefit from tactile demonstrations can be served by 3D prints, and the science that uses that mathematics can benefit similarly. (For example, in Chapter 2, we show how to use some of these surfaces to visualize light waves.)

We intended these models for surfaces that are difficult to visualize and that can be hard to think about in projected form. Sometimes, though, the most basic concepts are the hardest to grasp, and then a physical model of a fundamental concept can be a lifesaver.

Science Fair Project Ideas

If you decide you want to have a 3D printed mathematical function as part of a project, you will discover that you think differently about a math function when you are considering how to display it in 3D versus using the same 2D illustration that everyone else uses. Consider, for example, printing out a model of a function that applies to your product and taking measurements off it with a pair of calipers to see whether it agrees with theory. See if the creation of the physical model and any issues you had while printing give insight into the problem.

Summary

In this chapter, you learned several different ways to create a 3D model of a mathematical function in the OpenSCAD 3D solid modeling program, either one-sided on a flat base or as a thin two-sided sheet. You also saw how to create a file with an external program to pass through OpenSCAD in case you are generating surfaces in other codes and would like 3D models of them. Finally, we gave you a quick introduction to 3D printing and a few pointers to places to see more math examples and some ways to use these models in teaching and science fairs.

CHAPTER 2

Light and Other Waves

When someone says "waves," the first thing that may come to mind is the kind of waves that crash onto a beach. Water waves (and waves in air) are complicated, because the waves are moving over complex terrain, experiencing friction and effects of the water's viscosity and a lot of other things. The equations that govern the realities of water waves and their interaction require that you know about a type of math called partial differential equations, which even scientists and engineers might not work with much until graduate school.

Fortunately, though, light waves and other electromagnetic radiation follow some rules that simplify a lot of things for some special cases. The geometries of even these simpler interacting wave examples can be complicated, though. In this chapter, we will develop some 3D printable models that represent some of these special cases to help you visualize wave interactions. Students at all levels have trouble visualizing the properties of light and magnetism because of their abstract nature. If you are a teacher, you may find some of these materials useful just to pass around while discussing some of these phenomena.

CHAPTER 2 LIGHT AND OTHER WAVES

This chapter assumes you are comfortable with trigonometric functions like sine, cosine, and tangent and their inverses (asin, acos, atan). If not, you can check out our book *Make: Trigonometry* (2023, Make: Community LLC) for background and a different take on some of the topics in this chapter.

This chapter explores topics that are typically taught at levels from high school through the research level, and the eager student will have a lot of places they can go exploring further. The authors would like to acknowledge the assistance of astronomer and interferometry expert Stephen Unwin in clarifying the physics underlying this chapter and suggesting good ways to frame some of the basics without resorting to calculus. Any errors, of course, remain ours.

MODELS USED IN THIS CHAPTER

This chapter uses five OpenSCAD models. For more on 3D printing and creating and using math models in OpenSCAD, check out Chapter 1. Select printable STL example files are included in the book's repository as well. The OpenSCAD models are as follows:

- `basicWaves.scad` creates models of waves. The model is shown in the chapter as Listing 2-1.

- `multipleWaves.scad` is a more complex example adding two sources, shown in the chapter as Listing 2-2. (Note that this has changed from the first edition of this book; in the first edition, the third dimension was intensity, not amplitude.)

- `oneSlit.scad` models how light from a single slit propagates, as shown in Listing 2-3.

- `inverseOneSlit.scad` creates a model of the negative space of oneSlit.scad, to show how this space forms an envelope for twoSlits.scad. It is in Listing 2-4.

- `twoSlits.scad` shows light interfering from two slits, as shown in the chapter as Listing 2-5.

All these models are based on `surfaceprint.scad` from Chapter 1 (Listing 1-2). In the chapter, we only show the equation being graphed (`f(x, y)`), any variables introduced to define the function, and the thickness (`t`) variable. All the rest is the same as Listing 1-2. We do however have the files listed here in the repository for convenience, including the part that is a repetition of Listing 1-2.

Note Everything in this chapter was generated with OpenSCAD for modeling and PrusaSlicer for controlling the printer. The demonstration prints illustrated in this chapter were done on a Prusa Mk3s+ printer, using PLA plastic, with 0.2 mm layer height and no support other than a brim (or attached skirt, if you prefer to think of it that way) to glue down the first layer.

Physics and Math Background

The first set of models in this chapter are two-dimensional waves in the x, y plane like waves in water. How high the waves are in the third dimension is called their *amplitude*. The distance between crests is called the *wavelength* of a wave. The units of these quantities depend a bit on whether we are talking about a wave moving through space, in which case wavelength will be in terms of length, like meters. It gets a little complicated when waves are propagating in both space and time.

Assume for this section that our models are snapshots (in time) of a wave propagating in a plane. We show their amplitude as they propagate in the x,y plane as the model's height in the z direction. In the case of a water wave, the height of the water wave would be the z value.

For a light wave, it is better to think of the height (value of z) as the brightness of the light at that x, y position at a snapshot in time. In the discussion of light waves through slits later in the chapter, we will calculate the square of the amplitude and display that as the z value. This quantity is usually called intensity. Its time average is what a detector (like a CCD camera) would detect in the case of light waves.

If you are interested in getting more into the math behind the concepts of adding waves, check out the "Making Waves" chapter of our book *Make: Trigonometry* (2023, Make: Community LLC).

Note We replace the function f(x, y) used in Chapter 1's surfaceprint.scad model (Listing 1-2) to generate the models in this chapter. The listings in this chapter are just fragments that describe a function (sometimes, several interrelated functions) to avoid repeating the same model over and over. However, the downloadable OpenSCAD files include the entire model in each file, incorporating the part here plus the rest of the Listing 1-2 material. Comments in the repository models and shown here may vary somewhat for clarity and context.

Principle of Superposition

Imagine that you are on a still pond and you start poking a stick in and out of the water. Ripples will spread across the pond. Then suppose a friend nearby started doing the same thing. The ripples going in multiple directions would in some cases add up (creating a double height ripple) and in others, subtract or cancel out.

In real life, the waves on the pond will die out and have other complex interactions, but we can get a lot of insight into many kinds of electromagnetic waves (like light and radio waves) by modeling the waves

CHAPTER 2   LIGHT AND OTHER WAVES

as simple sine and cosine waves that interact with each other like the ripples you just imagined on the pond. (We are ignoring many effects like polarization, but we must start somewhere.)

For many kinds of waves, you can add the effects of the different waves, called the principle of *superposition* (see the Wikipedia article "Superposition principle"). When the waves add, it is called *constructive interference*; when they cancel each other out, it is *destructive interference*. We will use this principle and an adaptation of the surface plotting model from Chapter 1 (Listing 1-2) to visualize some classic experiments.

Basic Examples

To give you some feel for how different types of wave geometries might play out, let's walk through a couple of examples. Our simple 3D printed model makes it easy to put together complex combinations of sinusoids and see the wave interactions. As discussed later in the chapter, we will be simplifying some of the physics behind some of the wave phenomena we model, but you can still gain some good insight from playing with these models.

Point Sources and Plane Waves

This section looks at *point sources*, which put energy into a system in a concentrated way—sort of like the stick in the water we talked about earlier or a small bright light in a big room. Energy from that source would then go out in all directions radially from the point. The other basic type of wave we talk about is a *plane wave*—a straight line wave across the whole plane, like waves far out at sea, which can be due to a source far away. Then we look at how these appear when they are superposed on each other.

In our example, the plane wave has four times the amplitude of the radial source and a wavelength five times as long. In Figure 2-1, you can see what a snapshot of this looks like when 3D printed. The waves going

CHAPTER 2   LIGHT AND OTHER WAVES

from left to right (i.e., traveling in the *x* direction) are bigger and farther apart than the short ripples from the point source on the left, which is centered at the point (50, 50).

The model calculates a value for the function of the combined waves at each point (x, y) on a triangular grid going from (0, 0) to (100, 100). However, many of the functions are in terms of the distance from a source to a particular point in space. We calculate that distance in the function r(), which calculates the distance from the source at (cx, cy) to the point (x, y). Figure 2-1 shows just the point source; Figure 2-2, just the plane wave; and Figure 2-3, the two superposed.

We can visualize this as waves on a pond that are set up by two sticks. One is near the center, sending out circular waves. The other is near the shore, a long way away, so that when it gets to the center of the pond, the waves are very nearly parallel. These are snapshots in time after the waves have been generated for a while and reached a steady state.

Figure 2-1. *Just the point source radiating from the center*

CHAPTER 2   LIGHT AND OTHER WAVES

Figure 2-2. *The plane wave alone*

Figure 2-3. *Plane wave and point source interaction*

CHAPTER 2 LIGHT AND OTHER WAVES

Wave Model Details

All the models in this chapter are based on surfaceprint.scad from Chapter 1 (Listing 1-2). In the listings in this chapter, we only show the f(x, y) equation being graphed, any variables introduced to define the function, and the thickness variable, t. All the rest of each of the models is the same as in Listing 1-2.

All models used range = [100, 100]. This means that the models are all 100 by 100 points and by default will print out 100 by 100 mm (about 4 inches square, with thickness depending on other parameters that we describe for each case). The origin is at the lower right-hand corner, and x and y are both always positive and go from 0 to 100. Some of the models add an offset to the z value to make printing possible (since the model will fill the space below z = f(x, y) to z = 0 when using the t = 0 option).

In most cases, we used res = 0.25, for points calculated every 0.25 mm. More points are usually better, but more points can make the OpenSCAD model take a very long time to render. This value seems to be a good trade-off that keeps the feature size big enough to be printed while preserving the important features. You will need to play with this a little as you develop models.

As we will talk about later in the chapter, we will not normally 3D print these models such that the x and y axes of the models are printed aligned with the x and y plane (parallel to the build platform) of the 3D printer. When we talk about x, y, and z in this chapter, we mean the coordinates *of the model*.

Finally, OpenSCAD uses *degrees* in its trigonometric functions. In most of our narrative, we talk about angles in *radians*, since that is how most textbooks present the problems, and because it works out conveniently in many more advanced topics. Radians are defined in terms of the radius of a circle 1 unit in radius. Since the circumference of a circle is the radius times 2 times the irrational number pi (Greek letter π), equal to roughly

CHAPTER 2 LIGHT AND OTHER WAVES

3.14159, this "unit circle" will have a radius of 2π. In other words, we go around 2π radians to go around the full 360 degrees of a circle with a radius of 1. Thus, to convert from degrees to radians requires multiplying by $2\pi/360$, or $\pi/180$.

Listing 2-1. Function for Superposition of Radial and Plane Wave (file basicWaves.scad)

```
//File basicwaves.scad
t = 3;
function planar(x, y) = 4 * sin((30 / PI) * y);
function radial(x, y) = cos((150 / PI) * r(x, y, 50, 50));
function f(x, y) = planar(x, y) + radial(x, y);
```

Two Interacting Sources

What happens if we have two interacting point sources at one edge of the plane we are modeling? The model for that is given in Listing 2-2, and the model we printed (file multipleWaves.scad) is in Figure 2-4. We assume that our two sources are perfectly in phase with each other (meaning that they both start their cosine waves at the same time), located at one edge of our print at x = cx1 and x = cx2.

In Figure 2-4, we have plotted a snapshot in time of the amplitude of the waves after the sources have been going for a while. Note though that if we were to put a screen that cut across our model at any given point (parallel to the bottom of Figure 2-4), we would see a pattern of bright and dark proportional to the average intensity of the wave. The average intensity will be proportional to the square of the amplitude.

Note that radial lines of zero intensity (known as *nodes*) appear along lines of constant angle radiating from a point between the two sources. These appear where the two in-phase sources cancel out. In the next section, we compare this model with light going through two thin slits, which arguably look like two point sources.

47

CHAPTER 2 LIGHT AND OTHER WAVES

In Listing 2-2, `lambda` is the wavelength (in mm), `r` is the distance (also in mm) from the center of each of the point sources to the point (`x, y`) on the surface, and `factor` is an arbitrary scaling factor to adjust how large the features are. (Note that in the first edition of this book, `multipleWaves.scad` computed the intensity of the interacting point sources; the listing here is computing the sum of the amplitudes of the two point sources at a moment in time.)

Figure 2-4. *Superposition of two point sources: amplitude snapshot in time*

Note To make sense of the models that follow, think about the fact that `lambda` = 4 means that we go through a full cycle (360 degrees) every 4 mm. Hence, you will see a lot of 360/lambda terms.

Listing 2-2. Function for Superposition of Two Point Sources (file multipleWaves.scad)

```
factor = 0.5;
lambda = 4;
cx1 = 50 - 8;
cx2 = 50 + 8;
function f(x, y) = factor * (
  cos((360 / lambda) * r(x, y, cx1, 0)) +
  cos((360 / lambda) * r(x, y, cx2, 0))
) + 5;
```

Diffraction and Interference

Next, let's build on these ideas to create some models of the phenomenon of *diffraction*. Waves diffract, changing their geometry and how they interact with each other when they encounter a physical barrier of some kind (see the Wikipedia article "Diffraction"). For example, what happens when a wave front passes through a wall with either one or two slits in it and continues through on the other side?

Light Through One Slit

Physicists typically talk about the *intensity* of light at a given spot, which is proportional to the square of the amplitude of the wave at that position and time. Let's say we have a light source (Figure 2-5) shining through a narrow slit in a wall, with a screen parallel to the wall on the opposite side. We would see a bright area on the screen corresponding to the center of the slit, fading away. Figuring out how bright (and thus the value of the intensity as we move away from the center of the slit) is complex and more math than we want to get into here, but the answer turns out to be straightforward.

CHAPTER 2   LIGHT AND OTHER WAVES

Our physicist friends tell us that the intensity of light through one slit of width w (`slit` in our models) is proportional to

$$\text{Intensity} \sim \text{sinc}^2\left(\frac{\pi w \sin(\theta)}{\lambda}\right)$$

Here, λ (the Greek letter lambda) is the wavelength, and θ (Greek letter theta) is the angle defined in Figure 2-5. It turns out that this *sinc(x)* function (pronounced "sink") comes up often when we talk about waves. It is defined as follows:

$$\text{sinc}(x) = \frac{\sin(x)}{x}, x \neq 0 \text{ and } \text{sinc}(0) = 1.$$

Since 0/0 is always "undefined" mathematically, we need to find an indirect way to get a value of *sinc*(0). It turns out that a calculus concept called *L'Hôpital's rule* (see our book *Make: Calculus*) will let you figure out that *sinc*(0) = 1. Note that some types of engineers define *sinc(x)* differently, so be sure you are using our definition. Our definition assumes that x is in radians.

CHAPTER 2 LIGHT AND OTHER WAVES

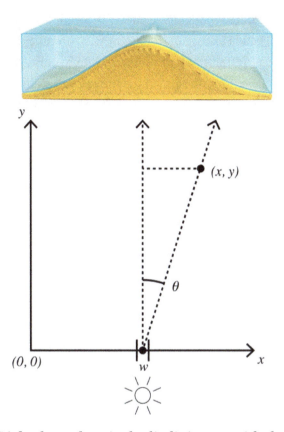

Figure 2-5. *Light through a single slit, lining up with the single-slit model*

We have 3D printed this intensity function with the model `oneSlit.scad` (Listing 2-3). How the bright (high) and dim (low) areas of our 3D printed model line up with the experiment's geometry is shown in Figure 2-5 (except that the wave's height would be in the z direction, out of the page). The angle θ shown in Figure 2-5 is the angle between a line to a given point (x, y) and a line parallel to the direction of propagation of the light wave through the slit.

51

We have also created a negative space version of this model, `inverseOneSlit.scad` (Listing 2-4), for comparison purposes we will get to shortly. The listing for these two models and one to come are all collected in the section "Diffraction Models," since they require some discussion that would be a distraction from the physics we are talking about here. The two are shown fitted together in the top part of Figure 2-5.

The Double-Slit Experiment

In 1803, Thomas Young performed an experiment using two narrow slits. This double-slit experiment was designed to quantify what the pattern of bright and dark areas would be if someone were to hold up a screen after the light had passed through these slits (see the Wikipedia article "Young's interference experiment" for more). A physics note: in this section, we talk about a monochromatic (one wavelength of simulated light) version of the experiment. The original experiment used multiple wavelengths and produced a rainbow interference pattern.

The setup of Young's double-slit experiment is like that of our single-slit setup, except that now the light from one source is passing through two slits. The two slits are assumed to be larger than the wavelength of the light passing through them, close together relative to the overall area we are modeling.

Young found that what he saw were alternating light and dark bands on a screen behind his slits, showing how the wave front going through each slit interferes constructively and destructively at different points in space. This experiment was taken as proof that light travels in waves that can be added together to produce interference patterns. He found that, for two

CHAPTER 2   LIGHT AND OTHER WAVES

slits of width w, a distance d apart (slit_separation in the models), the intensity at a given point as a function of angle away from the centerline is proportional to

$$Intensity \sim sinc^2\left(\frac{\pi w \sin(\theta)}{\lambda}\right) cos^2\left(\frac{\pi d \sin(\theta)}{\lambda}\right)$$

where the other variables are as we described them for the one-slit case. This works out to the two-slit intensity being the intensity pattern from one slit multiplied by a (squared) cosine wave. These equations can be derived by using *Fourier Transforms*, which are an engineering technique you will learn if you study electrical engineering or mathematical physics. But let's see what intuition we can get from these relationships without trying to derive them. The models in Figure 2-6 represent the one- and two-slit equations for intensity. We created a "negative space" version of the one-slit case (left) so we could explore its relationship with the two-slit case (right).

Figure 2-6. *One-slit intensity function (left) and the negative-space version over the two-slit one*

Figure 2-7 gives a top view of the one-slit intensity pattern (inverted, top left; positive, bottom left) and the double slit (top right). In all cases, they were printed with t = 0 (flat bottoms). The equations used are far-field equations (known as *Fraunhofer diffraction*) and do not really apply

CHAPTER 2   LIGHT AND OTHER WAVES

right at the slits. You see some nonphysical artifacts near the slit if you look closely. The interesting part is how the one-slit intensity curve will enclose the two-slit one. Let's explore the implications of these two equations:

$$Intensity, one\ slit \sim sinc^2\left(\frac{\pi w sin(\theta)}{\lambda}\right)$$

$$Intensity, two\ slits \sim sinc^2\left(\frac{\pi w\ sin(\theta)}{\lambda}\right) cos^2\left(\frac{\pi d\ sin(\theta)}{\lambda}\right)$$

For the two-slit case, if the slits are small (w near 0), we more or less are computing $sinc(0) = 1$, and the cosine term dominates the shape. But if the slit width(w) is large, it will sort of "smear out" the cosine pattern. Or to put it another way, slits that are not infinitely thin may start to wash out the fringe pattern.

You can think of each slit as a lot of smaller slits all shining together. In the straight ahead direction, that has little effect, as the light paths are all pretty much the same. But as you go off-axis (off the centerline, increasing θ), the waves do not line up nicely but tend to cancel out. The fringes are still there, but their amplitude is reduced (right side of Figure 2-7).

CHAPTER 2   LIGHT AND OTHER WAVES

Figure 2-7. *Inverted one slit (top L), two slits (top R), one slit (bottom L)*

Conversely, if the slits are very close together (d very small), then the cosine term trends toward 1, and the *sinc* squared factor dominates the behavior. Both how far apart the slits are placed and how wide each one is will determine how many waves "fit" into the "envelope" (roughly three in this example).

Note that you will not see a curve shaped like either of these if you do the experiment. In the case of a single slit, if we were to put a screen behind the slit, the height of the curve in our model is how bright a spot at the appropriate angle to the source would be, relative to other spots at that distance from the slit(s).

Likewise, for the two slits, our models tell us where bright and dark lines parallel to the slit would appear on the screen, called *fringes*. Applying this phenomenon for useful purposes is called *interferometry*. It has many applications in fields ranging from astronomy to manufacturing; we discuss some of them in the "Where to Learn More" section later in this chapter.

CHAPTER 2 LIGHT AND OTHER WAVES

Finally, go back and look at our two point sources superposed, back in Figure 2-4. You can see how the 2D model of two coherent point sources is a 2D analog to start understanding how light passing through two slits will interfere, with bands of dark and light being created by the interference if you imagine a screen inserted at right angles to the model.

Note For printability, the model in Figure 2-4 needed to have a wavelength only a little smaller than the distance between the sources, which blurs out the radial lines of dark and light a bit. The wavelength in all these models is close to the size of the slit and distance between them, again for printability, which violates some of the assumptions we have already noted. But we can get qualitative understanding just the same.

Diffraction Models

OpenSCAD functions for each of the objects shown in Figure 2-7 are in Listing 2-3 (one slit), Listing 2-4 (one-slit negative space), and Listing 2-5 (two slits). In those listings, we have a function sintheta(x,y). This function computes the sine of the angle theta (θ) from the geometry shown in Figure 2-5. Here we calculate $sin(\theta)$ directly as opposite over hypotenuse, as shown in Figure 2-5, rather than by calling the $sin()$ OpenSCAD function. Note that these models can take a while to render in OpenSCAD; if they take too long, make the res value bigger, around 0.25. We lowered it to 0.1 in the two-slit model to get more structure.

Note The pow function in OpenSCAD raises a variable to a specified power, so pow(x, 2) would return *x* squared.

CHAPTER 2 LIGHT AND OTHER WAVES

Listing 2-3. Model for the One-Slit Case (file oneSlit.scad)

```
//File oneSlit.scad
//wavelength, same units as slit
lambda = 4;
//width of slit
slit = 8;
//center of slit
c = 50;
//scaling factor, for visibility
factor = 20;
function sinetheta(x, y) = (x - c) / sqrt((x - c)^2 + y^2);
function slit_sinc(x, y) = lambda *
  sin(180 * sinetheta(x, y) * slit / lambda) /
  (PI * slit * sinetheta(x, y));
function f(x,y) = factor * pow(slit_sinc(x, y), 2) + 2;
//This function is now the SINGLE-slit experiment.
//sinc squared of (slit * pi * sin(theta)/lambda)
```

Listing 2-4. Model for the "Empty Space" Inverse of the One-Slit Case (file inverseOneSlit.scad)

```
//File inverseOneSlit.scad
//wavelength, same units as slit
lambda = 4;
//width of slit
slit = 8;
//center of slit
c = 50;
//scaling factor, for visibility
factor = 20;
function sinetheta(x, y) = (x - c) / sqrt((x - c)^2 + y^2);
function slit_sinc(x, y) = lambda *
```

57

CHAPTER 2 LIGHT AND OTHER WAVES

```
   sin(180 * sinetheta(x, y) * slit / lambda) /
   (PI * slit * sinetheta(x, y));
// This function is now the SINGLE-slit experiment, inverted
// (to show how other cases fit into it)
function f(x,y) = 24 - factor * pow(slit_sinc(x, y), 2);
// sinc squared of (slit * pi * sin(theta)/lambda)
```

Listing 2-5. Model for Two Slits (file twoSlits.scad)

```
//File twoSlits.scad
res = .1;
//wavelength, same units as slit
lambda = 4;
//width of slit
slit = 8;
//center of slit
c = 50;
//distance between slits
slit_separation = lambda * 4;
//scaling factor, for visibility
factor = 20;
function sintheta(x, y) = (x - c) / sqrt((x - c)^2 + y^2);
function slit_sinc(x, y) = lambda *
   sin(180 * sintheta(x, y) * slit / lambda) /
   (PI * slit * sintheta(x, y));
//note PI/PI goes away inside sin
function slit_cos(x, y) = cos(180 * sintheta(x, y) *
   slit_separation / lambda);
function f(x, y) = factor * pow(slit_sinc(x, y), 2) *
   pow(slit_cos(x, y), 2) + 2;
//This function is now the DUAL-slit experiment.
//sinc squared of (slit * pi * sin(theta)/lambda)
```

Caution In these models, we cheated a little by avoiding situations where we would need to compute sinc(0), which is defined as equal to 1. (It is 0/0 otherwise.)

Printing Considerations

All the objects in this chapter have been printed standing on end. Depending on the orientation that will have the fewest overhangs, you may want to rotate the object in x or y plus or minus 90 degrees to get to this point.

The other advantage of printing these functions vertically is that fine detail will print much better. If we have details with a lot of gaps between them, like the two point sources example (back in Figure 2-2), we will get a lot of stringing if you try to print a layer with a lot of gaps. (Stringing is the nozzle drooling fine hairs between parts when it jumps over gaps between one model section and another within a layer.) But printing it vertically means that each layer essentially has no gaps on the surface perimeter and just internal ones for the infill layers.

Extremely thin features, like the sharp end of the `oneSlit` and `twoSlit` examples, may still lose a bit of detail, both while generating the model and while slicing it, because these features are too thin to be represented by the model's resolution, the width of the printer's nozzle, or the height of the print layers (depending on orientation).

The print will stick well if you use a brim, which is essentially a few extra outlines around the first layer, which you need to peel off after printing. You may want to have a wider and slower first-layer extrusion to make the print stick even more.

CHAPTER 2   LIGHT AND OTHER WAVES

THINKING ABOUT THESE MODELS: LEARNING LIKE A MAKER

When we started working on this chapter, we found ourselves asking a lot of fundamental questions about what we were trying to represent and which models would give the most insight. One of the challenges here was that waves normally propagate in space *and* time, so in some ways, a 3D model has the same issues as a 2D video simulation. You have three dimensions, but either it must be a snapshot in time or one of the spatial axes needs to represent something other than the third dimension. Thinking about that and when it is worthwhile to have a "two-sided" model took a lot of time and discussion with several physicists.

In the "Science Fair Project Ideas" section later in this chapter, we talk about magnetism and why we decided not to include any models of magnetic fields. Magnetic fields were more difficult to think about in many ways because the magnetic fields have both a magnitude and a direction, which is difficult to show in a 3D static model. We explored models of simple bar magnets, as well as those of Earth's magnetic field and of the Sun. However, very few simple equation solutions give very much insight, and many field equations are in very inconvenient (for 3D printing) polar coordinate systems. Others are designed for a static simple case that gives one number—not a very interesting 3D print.

A little digging revealed that as a practical matter, many people who need to model these things rely on one of several semi-experimental models. It seemed to us like there were not a lot of models for which a surface without directional markings would be an improvement to a 2D drawing with arrows showing the field direction. We talk about this a bit more in the next section.

One of the things that came up with the set of models for this chapter was the idea of an *envelope model*. You could see easily that the two-slit pattern fit inside the envelope created by the one-slit case but creating a "negative space" model so that the two could be fit together was interesting. We encourage you to think about other applications of this and places where it might give you more insight. Basically, all you do is subtract the function of interest from a number big enough so that the negative space "fits inside" an outer box; see Figures 2-6 and 2-7.

Because we elected not to use any math here beyond high school–level geometry and trigonometry, some of the discussion of the amplitude of the waves or why you may want to look at the square of the amplitude for a time average is of necessity a little vague and ignores some dependencies on time, frequency, and other things that are important in real applications. If you need to go beyond these rough conceptual ideas, be sure you know the equations governing your case and when the simplistic approach here is appropriate and when it is not.

Where to Learn More

A good general background on the topics in this chapter can be found in the Khan Academy's (www.khanacademy.org) discussion of light waves in its physics section and many articles on Wikipedia. A good introductory college-level physics or electrical engineering textbook will help for the next level of exploration beyond what we have done here.

As we noted earlier, we get into more of the math behind the concepts of adding waves and other phenomena like reflection and refraction in the "Making Waves" chapter of our book *Make: Trigonometry* (2023, Make: Community LLC).

The material in this chapter could be taken in many different directions. We have not even touched on the many electrical engineering applications of *Fourier Transforms*, a technique that uses the mathematics of combining waves to solve engineering problems (see Wikipedia, "Fourier Transform").

CHAPTER 2   LIGHT AND OTHER WAVES

The properties of interfering waves are used routinely in applications that require a lot of precision. Astronomers use *radio interferometry* to create images of objects that, as they would say, subtend a small angle in the sky. This can either be something small that is sort of far away or something big that is very, very far away. Astronomers make interferometry observations by taking signals from two radio telescopes (those big dishes like those seen in the movie *Contact*, for instance) and thinking of the two telescopes as two slits admitting light from a point source very far away (in this case, it might be another galaxy).

The farther away the telescopes are from each other, the finer the detail of the object that can be taken from the radio waves. If there are groups of telescopes at different distances from each other, images can be built up from the signals of different resolutions of the same object. The VLA (Very Large Array, `www.vla.nrao.edu`) in Socorro, New Mexico, was designed for this application with 27 radio telescopes arranged in a Y-shaped configuration.

A nearer-to-home application is laser interferometry for precise measurement of optical and other precision surfaces that are being machined. Laser beams interfere with each other when the surface is shaped correctly, and so the pattern from a test beam and one reflecting off the surface will give information about surface inaccuracies.

As mentioned at the beginning of the chapter, the full equations that govern the wave motions of light are complex partial differential equations. Light waves are part of the broader category of electromagnetic waves. The physics of light waves and magnetism are tied together with *Maxwell's equations* (see Wikipedia article of that name), which do not lend themselves to simple solutions in OpenSCAD.

Magnetic fields have an additional representational problem in that they are directional. The simplest example of this is a bar magnet, with its north and south poles. Typically, people show magnetic field lines by sprinkling iron filings around a magnet and then noting the north and south poles of the magnet. This seemed to us to be a case where adding a third dimension did not add a lot to the discussion, since a two-dimensional slice through a field with directional arrows in some ways is more information than a three-dimensional surface might be.

People who have to model Earth's magnetic field for navigational or other purposes resort to one of several semi-experimental models that are available. The Earth's field is "blown back" heavily by the solar wind—charged particles released from the Sun. The solar wind varies over time, and this variation is known as space weather.

The Sun's own magnetic field has some interesting interactions and has semi-experimental models. One commonly used model is based fundamentally on an ancient mathematical construct called an *Archimedes spiral*. This model (see the Wikipedia article "Heliospheric current sheet") is referred to as the *Parker spiral* after its originator.

We created some simple models that looked like magnetic fields, but after some debate, we decided that the models were too simplistic and could lead readers to draw some incorrect conclusions about the more-complete mathematics. Making them "more right" would have taken us places that we felt were beyond the intended audience for this book and the desire to use simple and open source tools wherever possible. Thus, we have decided not to include any magnetism models here, but we encourage advanced students and their teachers to think about what magnetic field line or potential surface models would be good for more insight. Exploring a special case or narrow application might lead to some interesting areas to explore.

Teacher Tips

Waves in fluids are taught at various points in the K-12 curriculum. Light waves are dealt with in a somewhat cursory way, since as noted earlier, many properties of waves cannot be fully appreciated until students have had some exposure to undergraduate-level physics and math.

CHAPTER 2   LIGHT AND OTHER WAVES

However, playing with these models to build some intuition can motivate learning about some of the precursor material or lead to science fair or extra-credit projects. That said, if you enter "waves" on the Next Generation Science Standards website (www.nextgenscience.org), you will get a lot of hits. Here are some we thought might be particularly good places to use the models here, or develop your own:

- www.nextgenscience.org/msps-wer-waves-electromagnetic-radiation
- www.nextgenscience.org/4w-waves
- www.nextgenscience.org/hs-ps4-3-waves-and-their-applications-technologies-information-transfer
- www.nextgenscience.org/4-ps4-1-waves-and-their-applications-technologies-information-transfer

As with Chapter 1, we hope this chapter is a starter set that will give you ideas for your own projects.

Science Fair Project Ideas

Precisely because these concepts rapidly lead into some sophisticated areas, we can imagine science fair projects that involve designing and printing a variety of different types of waves to compare, contrast, and measure. It seems to us that some interesting exercises could be built around one function being the envelope or negative space of another. The simple superposition models are particularly suited to compare to models in a *ripple tank*, using water waves.

CHAPTER 2   LIGHT AND OTHER WAVES

More Wave Interaction Models

If you know that a complex function can be approximated by summing sine waves of different frequency and amplitude, you can experiment a bit in OpenSCAD and then print out the ones that look particularly interesting or that you would like to try comparing or fitting into one another. For rapidly variable functions, it might get challenging to print steeply sided curve. The way the model is designed, you can put in more points, though they will still be 1 mm apart (but the piece will get bigger). You can scale a 3D print in PrusaSlicer or other 3D printer slicing software if the piece gets too big for your printer, although then you might once again have overly steep vertical sides to print.

Summary

This chapter developed some 3D printable models of interacting sinusoidal waves, which are useful for modeling the physics of light, magnetism, and other phenomena. We learned about what happens when some different types of waves interact and how a light wave behaves after it passes through one or two thin slits. We learned some 3D printing techniques (such as printing a thin object with a lot of detail on its side) and some modeling techniques (combining "negative space" models with conventional positive ones). Finally, we gave various ideas for topics that could be taught with these models at grade levels from middle through graduate school.

CHAPTER 3

Gravity

No one questions the existence of gravity in everyday life. When we put something on a table (astronauts excepted), typically we expect it to stay there. Gravity is mathematically a little subtle, though, and in this chapter, we look at the gravity field around the Earth and Moon, as well as that inside the trinary system that we see as the star Algol.

We also look at how moons or comets speed up or slow down in their orbits, as Johannes Kepler figured out between about 1605 and 1618. A few decades later, in the 1660s, Newton developed calculus to be able to better calculate planetary orbits. You will not need any calculus to understand this chapter, but you should be comfortable with a bit of trigonometry and some physics and astronomy. We will define terminology as we go and give some links in the main text as well as in the "Where to Learn More" section at the end.

CHAPTER 3   GRAVITY

MODELS USED IN THIS CHAPTER

This chapter uses two different OpenSCAD models. For more on creating and using math models in OpenSCAD and the basics of 3D printing, check out Chapter 1. Select 3D printable STL example files are included in the repository as well. The OpenSCAD models are

- `gravity.scad`: This model produces the gravitational potential surface of the Earth-Moon and Algol systems as described in the chapter. It is included as Listing 3-1.

- `orbitalVelocity.scad`: This model produces models of the orbits of planets (and their moons) and comets, with the third dimension being the speed of the planet at that point in its orbit. It is included as Listing 3-3.

Universal Gravitation

What is gravity? There have been many explanations throughout history, some mechanistic and involving "aether particles" and others more exotic. There was a big argument for about a century, starting with Newton and Leibnitz in the mid-1600s, about whether there was a *vis viva*, or life force, that somehow caused the effects of gravity, with an "aether" mediating it.

Newton thought the *vis-viva* idea was nonsense, and the only place the term survives today is in his equation of planetary orbit velocity (often called the *vis-viva equation*), which we model later in this chapter. People are still trying to find Grand Unified Theories that will tie gravity and other forces together, but there is not one as of this writing. Einstein's general relativity (see "Relativity" in Wikipedia) is the currently accepted framework for understanding gravity theoretically—and incidentally, as we were writing the first edition of this book in 2016, gravitational waves

CHAPTER 3 GRAVITY

predicted by his explanation were observed for the first time. If you are interested, there is a (much simplified) gravitational wave model in our book, *3D Printed Science Projects Volume 2* (Apress, 2018).

In this chapter, we look at ways to visualize the gravity of planets and stars and how stars, planets, and moons orbit each other. Kepler and Newton figured a lot of it out 500 years ago, and the basics let us think about why the solar system is like it is and why it is a good idea to build bases on the Moon to explore the solar system.

Newton's law of universal gravitation, which he laid out in his book *Philosophiæ Naturalis Principia Mathematica* (usually just called *Principia*) in 1687, said that the gravitational force between any two bodies is proportional to the two masses (m_1 and m_2) multiplied together and divided by the square of the distance (r) between them. The law is usually written like this:

$$Force = \frac{G m_1 m_2}{r^2}$$

where G is the *universal gravitational constant* (6.674×10^{-11} m^3/kg-s^2) and r is the distance between the two masses m_1 and m_2. These forces are *vectors*, which means they have a direction associated with them as well as a magnitude, so you must be careful to figure out how the forces add when a bunch of bodies are involved.

On Earth, gravity always points down toward the center of the Earth. Even Mt. Everest is a small enough bump on the surface of the Earth that at its peak you are not significantly farther enough from the center of the Earth to feel any difference. Even at the altitude of the space station, the difference is only about 11%, which we can calculate by comparing the force calculated at a radius of 6371 km at the surface of the Earth versus the force 400 km farther out, at the space station orbit. (People are weightless in orbit around the Earth because they are essentially always falling, which is a different effect.) What happens, though, if you want to know about how the effort to overcome gravity varies as you move farther away from a planet or star?

CHAPTER 3   GRAVITY

Suppose we want to simplify this a bit to get a little intuition about what is going on. In those cases, rather than the *gravitational forces*, we can look at the *gravitational potential*. This admittedly abstract quantity is best thought of as a difference: to move around from one place to another, we have to add potential energy or possibly get energy back, like we do if we raise or lower a bucket of water on Earth. It can be a little trickier, though, because to get from one point in space to another, you might have to "climb up" higher in between the two points and then get that energy back farther along your path.

The gravitational potential adds up all the forces from various bodies in the system we are studying and gets a single number (a *scalar*) for any particular point in space and time. The potential at any point for a particular unit of mass works out to be proportional to the contribution of other masses around it, each divided by the distance to that mass. In other words, each mass adds to the work that needs to be done to get away from that mass in a way that gets bigger with the mass getting larger and gets smaller as the mass gets farther away.

$$Potential = Gm_1\left(\frac{m_2}{r_2} + \frac{m_3}{r_3} + \ldots\right)$$

Note that the potential drops off as 1/distance, whereas the force drops off as 1/(square of the distance). In calculus-speak, the potential at a point is the integral of the forces. But if we imagine all the masses we are paying attention to are concentrated into a point at their respective centers (*point masses*), we can just think of it as adding up the influences of the masses in the system, adjusting for distance. If we do not assume we are dealing with point masses, we need to wade into some calculus (in fact, this problem was one of the reasons Newton invented it). But let's see how far we can go with this simple equation.

CHAPTER 3 GRAVITY

Gravitational Potential Wells

A planet (or moon) is a big mass, and so to get away from it, you can think of "climbing out" of its gravity. The good news is that the amount of energy required to climb the next increment falls as you get farther away. The equation leads to a characteristic shape that we will see in a minute. As a side note, one of the better 2D graphics of this was done in the science cartoon XKCD, at https://xkcd.com/681/. A little more conventional explanation can be found at the Wikipedia article "Gravitational potential." Let's explore a 3D printed model of the physics.

Earth-Moon System Model

Think of the Earth and the Moon as two points, each with the appropriate amount of mass concentrated at that point. Imagine a 2D plane containing both points, which we will call the x-y plane. Now, let's create a surface (in the z dimension, perpendicular to this x-y plane), the height of which is the gravitational potential due to the sum of the Earth's and the Moon's gravity at each point in that plane, based on the equation in the previous section. In Figure 3-1, we show an OpenSCAD 3D plot of the resulting summed gravitational potential surface. We are plotting the potential as negative values, so we can think of this as energy that will be taken away from our rocket or other energy source as we "climb out of the well" around a planet.

It takes a lot of energy to pull something from the surface of the Earth to a place between the Earth and Moon. However, the Moon is a lot smaller than the Earth (about one-quarter the diameter and less dense), and so as it turns out, its contribution is much less. The long spikes in Figure 3-1 that go to infinity from either body are attempting to show that there is an undefined point right at the center of the Earth and Moon, which could each be represented as a vertical pole through the center of the wide and narrow "wells," respectively.

CHAPTER 3   GRAVITY

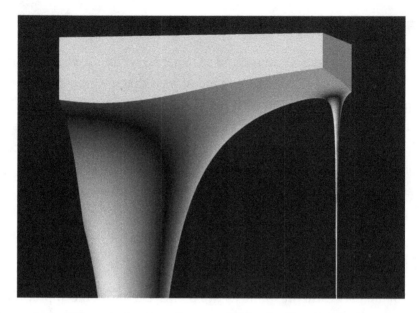

Figure 3-1. *The gravitational potential in the Earth-Moon system (ignoring all other bodies)*

We would like to 3D print a model of this so we can handle it and get some intuition. Since as you can see from Figure 3-1 the peaks would not be printable, we have instead *subtracted* these values from a constant and 3D printed the resulting surface (Figure 3-2), of course truncating the functions that go as $1/r$ as r approaches zero. The surface of the Earth or Moon is a finite distance out from the center, so we would cut off the singularity that way as well.

Remember that the height of the curve in Figure 3-1 and of the 3D print in Figure 3-2 are representing an energy value, not a physical distance. Imagine a plane parallel to the base of the model. In this simplified universe, Earth would be a dot lined up with the center of the big well on the left of Figure 3-2, and the Moon would be a dot at the center of the skinny well. The bigger curve that dominates the 3D print is (part of) the Earth's well, which is shaped similarly to the Moon's.

CHAPTER 3   GRAVITY

Figure 3-2. *3D print of Earth-Moon system gravity potential surface*

Let's call this plane parallel to the base the *x-y* plane and the height of the surface *z*. As we move around this two-dimensional plane with the centers of the Earth and Moon on it, the difference in height between one point on the surface and another would represent how much energy we would need to expend to move between the corresponding *x-y* points.

To put it another way, the front face of the model in Figure 3-2 is a graph of the potential as we move along that line from left to right, passing first through the center of the Earth and then through that of the Moon. As we move away from us on the 3D surface, moving parallel to that front face, we are graphing the potential along lines parallel to that Earth-Moon line. We get the curved surface we can see reflecting the contribution of both the Earth and the (much smaller) Moon.

People think of "climbing" to a higher potential energy out of a "well" centered on a body. We can think of this surface as the potential energy of the static Earth-Moon system, if somehow for an instant everything stopped moving. A difference in height on the surface is a difference in

CHAPTER 3 GRAVITY

potential energy, like a bucket of water being higher or lower over a well. For example, it would be better to build a supply depot on the Moon if you were creating bases in deep space rather than having to lug up everything from the Earth's well. As a side note, the instantaneous *slope* of any point on this surface is the net gravitational force being exerted on a body as it moves around the Earth-Moon system.

The Earth is nearly 100 times the mass of the Moon, so to see the gravity well of the Moon at all, we can only show a small part of the well of the Earth. Listing 3-1 is the model (file earthMoon.scad) that created the 3D print in Figure 3-2. We printed the models hollow, using "vase mode." See the "Printing Tips" section of this chapter for how that works. The model is based on the simple math surface models in Chapter 1, graphing the gravitational potential in the z direction for each point in x and y. To print the Earth-Moon system, select EarthMoon in OpenSCAD's Customizer. The rest of the variables will be set for you.

Caution For all the models in this chapter, the scaling in any dimension is arbitrary if we are consistent. The "real" numbers are very big, so we have used scaling factors liberally to make features visible. The center of the Earth and Moon are "singularities" and would require infinite energy in this simplified example.

Listing 3-1. Gravitational Potentials (file gravity.scad)

```
//OpenSCAD model to compute gravitational potential wells
//File gravity.scad
//(c) 2016-2024 Rich Cameron
//for the book 3D Printed Science projects, Volume 1
//Based on triangleMeshSurface.scad
//Licensed under a Creative Commons, Attribution,
```

CHAPTER 3 GRAVITY

```
//CC-BY 4.0 international license, per
//https://creativecommons.org/licenses/by/4.0/

//Attribute to Rich Cameron, at
//repository github.com/whosawhatsis/3DP-Science-Projects

//Thickness along z axis. t = 0 gives a flat base at z = 0
t = 0;
//Range of [x, y] values to graph
range = [50, 100];
//resolution in mm (smaller = smoother, but slower render)
res = .25;
blockymode = false;
//which system to represent
system = "EarthMoon"; //["EarthMoon", "Algol"]
//EarthMoon is a 2-body system, and Algol is a 3-body system.
//To represent a different system, you can modify the values
//below for one of these cases, or modify the f(x, y) function.

//z height of printed model, in mm
model_height = 100;
AU_to_mm = .036;
//AU_to_mm = .001;
zscale = 4; //here just a scaling factor in Z

mass_earth = 5970; //in 10^21 kg
mass_moon = 72; //in 10^21 kg

mass_Aa1 = 4.5; //masses relative to Aa2
mass_Aa2 = 1;
mass_Ab = 2.5;
a1_a2 = 0.062; //distance in AU
a1_b = 2.69; //distance in AU (approximate)
//This model assumes that the three bodies are aligned
//with Aa2 between Aa1 and Ab.
```

CHAPTER 3   GRAVITY

```
function f(x, y) = (system == "EarthMoon") ?
  max(0, model_height - .2 * (
    mass_earth /
      sqrt(pow(x, 2) + pow(y - 82.3, 2)) +
    mass_moon /
      sqrt(pow(x, 2) + pow(y - 5.5, 2))
  ))
: (system == "Algol") ?
  max(0,  model_height - zscale * (
    mass_Aa1 /
      sqrt(pow(x, 2) + pow(y - 15, 2)) +
    mass_Aa2 /
      sqrt(pow(x, 2) + pow(y - 15 - a1_a2 / AU_to_mm, 2)) +
    mass_Ab /
      sqrt(pow(x, 2) + pow(y - 15 - a1_b / AU_to_mm, 2))
  ))
: "Undefined system!";

assert(is_num(f(0, 0)), "Undefined system!");

s = [
  round((range[0] - res/2) / res),
  round(range[1] / res * 2 / sqrt(3))
];
seg = [range[0] / (s[0] - .5), range[1] / s[1]];

function r(x, y, cx = range[0]/2, cy = range[1]/2) =
  sqrt(pow(cx - x, 2) + pow(cy - y, 2));
function theta(x, y, cx = range[0]/2, cy = range[1]/2) =
  atan2((cy - y), (cx - x));
function zeronan(n) = (n == n) ? n : 0;

points = concat(
```

```
  [for(y = [0:s[1]], x = [0:s[0]]) [
    seg[0] * min(max(x - (y % 2) * .5, 0), s[0] - .5),
    seg[1] * y,
    zeronan(
      f(
        seg[0] * min(max(x - (y % 2) * .5, 0), s[0] - .5),
        seg[1] * y
      )
    )
  ]], [for(y = [0:s[1]], x = [0:s[0]]) [
    seg[0] * min(max(x - (y % 2) * .5, 0), s[0] - .5),
    seg[1] * y,
    t ? zeronan(
      f(
        seg[0] * min(max(x - (y % 2) * .5, 0), s[0] - .5),
        seg[1] * y
      )
    ) - t : 0
  ]]
);
*for(i = points) translate(i) cube(.1, center = true);

function order(point, reverse) = [
  for(i = [0:2]) point[reverse ? 2 - i : i]
];
function mirror(points, offset) = [
  for(i = [0, 1], point = points)
    order(
      point + (i ? [0, 0, 0] : [offset, offset, offset]),
      i
    )
];
```

CHAPTER 3   GRAVITY

```
polys = concat(
  mirror(concat([
    for(x = [0:s[0] - 1], y = [0:s[1] - 1]) [
      x + (s[0] + 1) * y,
      x + 1 + (s[0] + 1) * y,
      x + 1 - (y % 2) + (s[0] + 1) * (y + 1)
    ]
  ], [
    for(x = [0:s[0] - 1], y = [0:s[1] - 1]) [
      x + (y % 2) + (s[0] + 1) * y,
      x + 1 + (s[0] + 1) * (y + 1),
      x + (s[0] + 1) * (y + 1)
    ]
  ]), len(points) / 2),
  mirror([for(x = [0:s[0] - 1], i = [0, 1]) order([
    x + (i ? 0 : 1 + len(points) / 2),
    x + 1,
    x + len(points) / 2
  ], i)], len(points) / 2 - s[0] - 1),
  mirror([for(y = [0:s[1] - 1], i = [0, 1]) order([
    y * (s[0] + 1) + (i ? 0 : (s[0] + 1) + len(points) / 2),
    y * (s[0] + 1) + (s[0] + 1),
    y * (s[0] + 1) + len(points) / 2
  ], 1 - i)], s[0])
);

if(blockymode)
  for(x = [0:res:range[0]], y = [0:res:range[1]])
    translate([x, y, 0]) cube([res, res, f(x, y)]);
else polyhedron(points, polys, convexity = 5);
```

CHAPTER 3 GRAVITY

Algol Model

The star Algol, in the constellation Perseus, is composed of three stars, called Aa1, Aa2, and Ab. Their masses, relative to Aa2's mass, are 4.5, 1, and 2.5, respectively. As seen from Earth, Aa1 and Aa2 form an eclipsing binary as they orbit each other 0.062 AU apart. The third star Ab is 2.69 AU from the others, on average. All we need for our model are relative separations, so we will not worry about the units if they are consistent.

Algol was known in ancient times as the "demon star" (see Wikipedia, "Algol") because it varied in brightness every few days. We know now that this is because Aa1 is far brighter than Aa2 or Ab, and the planes of their orbits line up with the Earth so that Aa1 gets eclipsed every 2.85 days by Aa2 and less often by Ab, which orbits much farther away.

Listing 3-1 can be altered (in the Customizer) to create two different gravity potential models. The first one, shown in Figure 3-3, creates a 3D printed model of all three stars. The close pair (on the right) are hard to distinguish from each other, just because of the resolution limitations on a 3D print.

Figure 3-3. *The three-star Algol print*

CHAPTER 3 GRAVITY

We zoom in on the model in OpenSCAD (Figure 3-4) for a better view. To create the 3D print shown in Figure 3-3, open the `gravity.scad` file, and in the OpenSCAD Customizer, select `Algol` for the `system` variable, and change the `model height` variable to 40.

Figure 3-4. *A close-up of the field in Figure 3-4 showing the two close-together stars in the three-star Algol model*

You can also do the same model but adjust scales to zoom in more closely to see the two close-together stars in more detail. Figure 3-5 shows the close binary pair, with the third star cut off at this scale. For this one, again use the `gravity.scad` file, and in the OpenSCAD Customizer, again select `Algol` for the `system` variable, and change the `model height` variable to 40. However, to zoom in, also change the variable `AU_to_mm` to `0.001`.

CHAPTER 3   GRAVITY

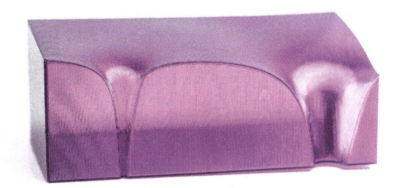

Figure 3-5. *The Algol gravity well print—a cross section of the field from Aa2 (left) and Aa1 (right)*

Custom Gravity Well Models

If you want to print out the gravity wells of two bodies not listed here, you can select the EarthMoon case in the Customizer and then, after doing that, change the masses of Earth and Moon to the relative masses of your two bodies:

mass_earth = 5970; //in 10^21 kg
mass_moon = 72; //in 10^21 kg

For three bodies, you can select the Algol case and edit the physical parameters similarly. What matters is the ratio of any masses or distances; the absolute numbers do not matter. You may need to tweak scaling factors to make the models printable.

Orbits

A gravity well is a useful concept if we want to think about the relative energy needed to get away from different planets or moons. However, planets do not hover, stationary, at a point in their star's gravity well. Instead, planets and stars *orbit* around one another in complex ways.

CHAPTER 3 GRAVITY

Since stars are heavy and planets are (usually) of almost negligible mass relatively speaking, typically this means that the star (in our case, the Sun) has the center of mass of the whole system of planets inside it. In a case like Algol (discussed in the preceding section), it is more complicated.

Instead, think about a simple case with one heavy body and one much lighter one orbiting it. Kepler figured out that in situations like this, the light body will travel in an *elliptical* orbit, with the heavy body at one *focus* of the ellipse (see the Wikipedia article "Ellipse"). Two bodies will orbit their mutual center of mass; in this simplified case, the center of mass of the system is assumed to be at the center of the larger mass.

Figure 3-6 shows the major features of an ellipse: the semimajor axis, usually referred to as *a* (half the longer diameter of the ellipse), and the semiminor axis, *b*, which is half the shortest diameter. The two foci lie along the major axis, at equal distances from the minor axis. If you draw two lines from any point on the ellipse to the two foci, the sum of the length of the lines will be equal to the length of the whole major axis (2a). The two foci are shown as dots on the major axis of Figure 3-6.

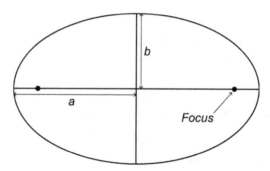

Figure 3-6. *Parts of an ellipse*

CHAPTER 3   GRAVITY

If we assume that the ellipse in Figure 3-6 is in the *x-y* plane, with the coordinate system origin at the center and the semimajor axis being along the *x* axis, the ellipse can be represented by the equation:

$$\frac{x^2}{a^2} + \frac{y^2}{b^2} = 1$$

Note that this becomes a circle if *a* and *b* are equal. We can relate the semimajor and semiminor axes of an ellipse to a quantity called the *eccentricity*, usually referred to as *e*. The eccentricity is zero for a perfect circle and approaches 1 for a long skinny ellipse. Its equation is

$$b = a\sqrt{1-e^2}$$

Thus, if we have the eccentricity *e* and semimajor axis *a*, we can get the semiminor axis, *b*.

Kepler developed three laws that govern the physics of orbits assuming, among other things, that they are ellipses. He published them between 1609 and 1619 (see Wikipedia, "Kepler's laws of planetary motion"). They are

- First law: The orbit of every planet is an ellipse with the Sun at one focus.

- Second law: A line from a planet to the Sun sweeps out the same area during equal amounts of time (which has the corollary that planets move faster near the focus containing the Sun).

CHAPTER 3   GRAVITY

The second law is illustrated in Figure 3-7. A string attached to the planet traveling in its elliptical orbit and the Sun would sweep out the red area and the blue (equal) area in the same amount of time.

- Third law: The square of the time it takes a planet to go completely around the sun is proportional to the cube of the semimajor axis (a) of the orbit.

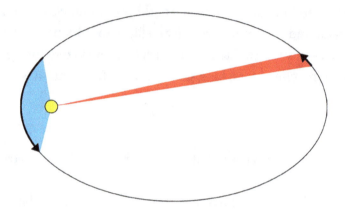

Figure 3-7. *Sweeping out equal areas in equal times*

Isaac Newton and others built on this about 66 years later to come up with the *vis-viva* equation (sometimes called the *conservation of energy* equation—see the Wikipedia article "Vis-viva equation"):

$$v^2 = GM\left(\frac{2}{r} - \frac{1}{a}\right)$$

where v = the orbital velocity (meters per second) when the orbiting body is at any point r (meters) in its orbit, G is the universal gravitational constant (6.674×10^{-11} m³/kg-s²), M is the mass of the central body (in kilograms), and a is the semimajor axis of the orbital ellipse in meters. The radius r is the distance between the central body's position at one focus of the ellipse and the orbiting body's position on the ellipse, which will vary as the orbiting body travels. You can read more about ellipses and a slightly different view of Kepler's laws in our book *Make: Geometry* (2021, Make: Community LLC), where Figures 3-6 and 3-7 first appeared.

In the case of the orbit models we create in this section, we assume that the Sun is so much heavier that the center of mass of the whole solar system is the mass of the Sun located at one focus of the ellipse and compute the orbits for each planet as a two-body problem with the Sun. In practice, it is way more complicated than that, but we can learn a lot from simplified models if we know their assumptions and limitations. To learn more, look up "two-body problem" and "three-body problem" in Wikipedia and elsewhere.

Halley's Comet Orbit Model

Figure 3-8 shows a model of an elliptical orbit with the height of the top surface being equal to the velocity an orbiting body would have at that point in the ellipse. In this case, it is the orbit of Halley's Comet around the Sun. Halley's Comet was the first comet to have its orbit worked out, in about 1705 by Edmund Halley (see the Wikipedia article "Halley's Comet"). The semimajor axis is 17.94 AU (astronomical units, the average distance from the Earth to the Sun).

The 3D print shown in Figure 3-8 has a little cone at the focus, where the Sun would be located, and a small hole at the other focus, near the very farthest point away from the Sun. They are hard to see in the Figure because the orbit is so stretched out that the foci look like they are almost falling on the ellipse.

The 3D print in Figure 3-8 is created with the model in Listing 3-2, file `orbits.scad`. In the case of this smaller Halley's Comet orbit model, use the `Customizer` to select `Halley` to set most of the parameters, and then change the variable `AU_to_mm` to 4 versus the default of 40, which scales the 3D print so that you can fit the whole orbit on a reasonably sized consumer printer. In the next section, we talk about how to create a partial comet

CHAPTER 3   GRAVITY

orbit model that is scaled the same as inner solar system orbit models. The height (velocity) of this model is, however, scaled the same as for the inner solar system models created in the next section.

Figure 3-8. *Halley's Comet orbit model. Height is instantaneous orbital velocity*

Listing 3-2. Modeling the Orbital Speed of Halley's Comet (file orbits.scad)

```
//OpenSCAD model to compute the velocity of a body in a
//Keplerian two-body-problem orbit
//File HalleysComet.scad
//(c) 2016-2024 Rich Cameron
//for the book 3D Printed Science projects, Volume 1
//Licensed under a Creative Commons, Attribution,
//CC-BY 4.0 international license, per
//https://creativecommons.org/licenses/by/4.0/
//Attribute to Rich Cameron, at
//repository github.com/whosawhatsis/3DP-Science-Projects

wall = 2;
//scaling factor for x and y axes
AU_to_mm = 40;
```

CHAPTER 3 GRAVITY

```
preset = 0; // [0:"",1:Mercury,2:Venus,3:Earth,4:Mars,5:Halley]
//degrees per segment (smaller = smoother, but slower render)
slice_angle = 1;
//semi-major axis in AU (if using "custom")
semimajor = 17.94;
//semi-minor axis in AU (if using "custom")
semiminor = 4.59;
//scale of z axis (arbitrary since z is velocity, not distance)
z_scale_factor = 25;
//keep this consistent between models to compare velocities

planets = [[semimajor, semiminor],
  [0.38709893, 0.3788], //Mercury
  [0.72333199, 0.7233], //Venus
  [1.00000011, 0.9999], //Earth
  [1.52366231, 1.5170], //Mars
  [17.94, 4.59]         //Halley's Comet
//Note that Halley is retrograde, so the speed relative to the
//relative velocity will be the sum, rather than the difference
];

size = AU_to_mm * planets[preset];

//reverse if semiminor > semimajor
a = max(size);
b = min(size);

echo(str("Long dimension is ", a * 2 + wall, " mm."));

$fs = .2;
$fa = 2;

//generate the wall using a chain hull
for(theta = [0:slice_angle:359.99]) hull()
```

CHAPTER 3   GRAVITY

```
  for(
    theta = [theta, theta + slice_angle],
    //generate an ellipse with one focus at (0, 0)
    x = a * (cos(theta)) + sqrt(pow(a, 2) - pow(b, 2)),
    y = b * sin(theta),
    r = sqrt(pow(x, 2) + pow(y, 2)),
    //Use the vis-viva equation to calculate the height
    //to represent instantaneous velocity
    h = z_scale_factor * sqrt(AU_to_mm) * sqrt((2 / r - 1 / a))
  )
    //generate a series of cuboids aligned along the wall
    translate([x, y, h / 2])
      //use tangent angle to the wall for uniform thickness
      rotate(atan2(a * y, pow(b, 2) * cos(theta)))
        cube([wall, .0001, h], center = true);

//generate the base
difference() {
  union() {
    translate([sqrt(pow(a, 2) - pow(b, 2)), 0, 0])
      scale([1, b / a, 1]) cylinder(r = a, h = 1);
    intersection() {
      //use a cone to represent the location of the sun
      cylinder(r1 = 5, r2 = 0, h = 5);
      translate([sqrt(pow(a, 2) - pow(b, 2)), 0, 0])
        scale([1, b / a, 1]) cylinder(r = a, h = 5);
    }
  }
  //make the sun cone hollow, to align stacked prints
  translate([0, 0, -1]) cylinder(r1 = 5, r2 = 0, h = 5);
  //use a hole to show the other focus
```

CHAPTER 3 GRAVITY

```
  translate([2 * sqrt(pow(a, 2) - pow(b, 2)), 0, 1])
    cylinder(r = 1, h = 10, center = true);
}
```

Halley's Comet has an eccentricity of 0.967, and this makes the semiminor axis 4.59 AU. Just to make things more interesting, Halley's moves around the Sun in a *retrograde* orbit, going around the Sun in the opposite direction than the Earth does, and its orbit is also tilted about 18 degrees relative to the plane of Earth's orbit.

When the comet is at its closest to the Sun (the big peak in the model in Figure 3-8), we can use the *vis-viva* equation with $r = 0.587$ AU (and the fact that 1 AU = 1.496×10^8 km). If we use kilometers and not meters as the length units for G for this problem, $G = 6.674 \times 10^{-20}$ km²/(kg - s²). Use 1.989×10^{30} kg for the mass of the Sun.

We wind up with the velocity of Halley's Comet at *perihelion* (closest point to the Sun) of about 55 km/s. This is remarkably close to the real value (usually quoted, just as we did, as "about 55 km/s"). The real value varies on the effects of other planets' gravity (notably that of Jupiter and Saturn).

To test out the equation for the Earth's speed around the Sun, we can, for purposes of calculation here, assume that the Earth's orbit is close to circular ($r = a = 1.487 \times 10^8$ km—we give the actual numbers later in Table 3-1). If you put those into the *vis-viva* equation, terms drop out, and you get out about 30 km per second. If you use the same radius of the orbit and say that the Earth goes around the Sun in an orbit that is about 2π times the same radius in a year ($365 \times 24 \times 60 \times 60$ seconds), gratifyingly we get the same number of about 30 km per second. (The official number is 29.8).

If you want to think about what would happen if Halley's Comet hit the Earth, the bottom line is that no one would be around to talk about it afterward. At $r = 1$ AU (where the comet's orbit could in principle cross Earth's orbit), the comet is going about 42 km/s. Halley's orbital inclination

CHAPTER 3   GRAVITY

relative to that of Earth means that it never actually crosses the plane of Earth's orbit when Halley's is at 1 AU, fortunately, so we do not have to worry about a collision.

Inner Solar System Model

Next, we will try printing the orbits of Mercury, Venus, and Earth to the same scale as each other. We will use the model in Listing 3-2, but with values as shown in Table 3-1. You can see how the three of them came out relative to each other in Figures 3-9 (side by side) and 3-10 (nested). For the planetary orbits, use OpenSCAD's Customizer to pick the relevant orbit. You should not have to change any other values.

The nested version is not completely accurate because each orbit has a 1 mm thick base, creating a 1 mm offset for each orbit you stack. But you can get a good qualitative idea of how these orbits relate to each other. As you would expect, the closer a planet is to the Sun, the faster it is going around in its orbit.

Table 3-1. Parameters Needed to Create an Inner Solar System Model

Planet	a (AU)Fi	b (AU)	e	AU_to_mm	Z_scale_factor
Mercury	0.3871	0.3788	0.2056	40	100
Venus	0.7233	0.7233	0.006773	40	100
Earth	1.00000	0.9999	0.001671	40	100
Mars	1.533	1.517	0.09341	40	100
Halley's Comet	17.94	4.549	0.9666	40 and 4	100

The stacked models in Figures 3-9 and 3-10 show this quite nicely, with Mercury's orbital eccentricity visible. Since the bases of the models are only 1 mm thick, they are translucent. Earth and Venus were so nearly circular that the two foci were both on the center point. If you are trying to be very accurate, you would need to look up how the orbits are phased. Although the Sun is at the same focus for both orbits, the semimajor axes might not be aligned.

The orbits also have some differences in how they are inclined relative to the Earth's plane, so they would be tilted a bit relative to each other, too (Mercury the most, at 7 degrees).

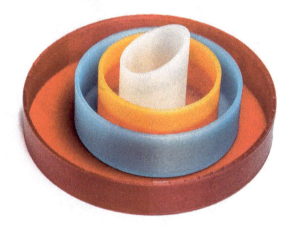

***Figure* 3-9.** *Mercury, Venus, Earth, and Mars orbits nested lined up on the focus (as if the Sun were at the focus of each orbit). The height of each point on the orbit is the instantaneous velocity*

CHAPTER 3   GRAVITY

Figure 3-10. *As in Figure 3-9 but seen from above. Note the offset from the common focus for Mercury and Mars*

We have also printed a version of the Halley's Comet orbital model scaled to be the same as the inner solar system models, shown in Figure 3-11 along with the Mercury orbital model. To make this model, again use the model `orbits.scad`, and select Halley. However, this time, leave the variable `AU to mm` at 40 instead of changing it to 4. In other words, scale it the same as the rest of the planetary orbits rather than making it smaller, as we did in the last section. The resulting model is to be too big to print on a consumer printer though. See the "Printing Tips" section of this chapter to see how to cut off just part of it.

Line up the cones showing the focus where the Sun is located so you can see the relative positions. Mercury is the only orbit that will fit entirely inside the Halley's Comet model. Figure 3-12 shows another view from the side. Figure 3-13 shows that the maximum height of the two scales of Halley models remains the same.

CHAPTER 3   GRAVITY

Figure 3-11. *Close-up of Mercury orbit model showing interior cone and hole at foci*

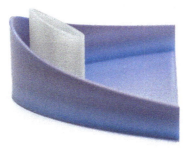

Figure 3-12. *Mercury's orbit and partial Halley's Comet orbit compared*

93

CHAPTER 3   GRAVITY

Figure 3-13. *Showing that the velocity variation is the same in the two scales of Halley's Comet models*

Note   In actuality, the orbit of Mercury is tilted 7 degrees from Earth's and Halley's is 18 degrees (technically, 180 - 18 degrees or 162 degrees, since it is retrograde), so these would not be lined up flat like this. But we ignore those effects here.

Custom Orbits

To create an orbit of a different planet, select the blank line in the preset pull-down menu, and then change the two variables, `semimajor` and `semiminor`, to your desired value. Tweak other scaling variables as needed.

Printing Tips

The models pictured in this chapter were printed on a Prusa Mk3s+, using PrusaSlicer software. The models in this chapter are relatively straightforward to print. The gravity well models were printed in *vase mode*, which means that they were printed hollow, with one end open (like

CHAPTER 3   GRAVITY

a vase). This was done both to make them photograph better and also to use less filament. The model in Figure 3-3, for example, was printed with the left-hand side down and the right-hand side open.

Vase mode can either be accomplished by selecting an option in your printer slicing software ("spiralize outer contour" in PrusaSlicer) or by manually selecting zero infill and zero top layers. (The latter is known as "old-school vase mode.") In Figure 3-14, we see the same print as in Figure 3-2, but now in the orientation in which it was printed. Note the thin wall and open top. This is a fast and accurate way to print models with a lot of side detail but no need for internal structure.

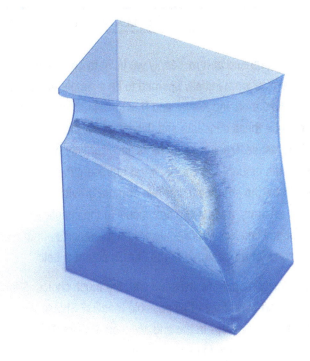

Figure 3-14. *The Earth-Moon gravity model, oriented as it was printed to show vase mode (open top)*

95

CHAPTER 3   GRAVITY

The orbits were printed with their large flat side down and are straightforward to print. One exception is that extremely elliptical orbits (like Halley's Comet) are a little challenging for older printers. The tallest parts are going to tend to get a bit too pointy to print completely cleanly. You can also add a tower in PrusaSlicer directly by right-clicking on the platform in the plater. Add a cylinder a bit taller than your model. This will give the plastic additional time to cool each layer, without the hot nozzle in contact with it. You can, of course, also accomplish this by printing two copies at once.

THINKING ABOUT THESE MODELS: LEARNING LIKE A MAKER

We continue to be surprised at how deeply we have to go into the physics background to develop what initially seemed to be a simple model. We have found ourselves reprising Kepler and Newton's lines of reasoning as we tried to represent these concepts and debating how closely to skate to subtle cosmology issues. Many planetariums have "gravity well" models, but it requires a little thought to consider what representations will actually give some insight for the interactions of two or more gravity wells. Scaling is fundamentally arbitrary, and for the first model in the chapter, we had to experiment some to come up with a scale where the Moon's effect was visible while retaining some of the features of the Earth's potential field.

For the orbital velocity model, we wrestled with too much choice. Should we try to show some sort of surface of all possible models? Should we pick one orbit, or try to interleave many? In the end, we went for simplicity, but you may want to experiment with ways to use these representations to do your own walk through the history of calculus and physics. We found ourselves wanting to do the experiment of printing a few planets at the same scale and looking at how the velocity distribution varied among the inner three planets. You could do a perspective graph in 2D, but it might not occur to you until you were playing with the models!

Where to Learn More

There are many places you can take these explorations. We cited Wikipedia background articles as we went along, but you can kick it up a notch by looking up data about other solar systems to make a gravity potential model or make a model of other planetary orbits or orbits of moons of planets. You might investigate the history of astronomy and particularly the history of the study of gravity. It seems hard to believe now, but gravity as a force without some medium to transmit it was very controversial for at least a century. You might think about how the models we made in this chapter might have looked different over the last two millennia.

The Kepler space mission (named after the Kepler we cited a lot in this chapter) has been discovering a lot of *exoplanets*—planets around other stars. To learn more about these systems of stars and planets beyond our own, check out https://science.nasa.gov/exoplanets/.

If you want to understand ellipses and how they are constructed a little better, you might want to look at the relevant chapters in our books *Make: Geometry* (2021) or *Make: Trigonometry* (2022), both published by Make: Community LLC.

Teacher Tips

Much of the material in this chapter lends itself to supporting undergraduate physics and/or calculus since some of the early development of calculus was in fact intended to solve these problems. The special cases (two-body problems) we show here are good for building intuition but need to be extended carefully to more general cases where these simple models may not apply (see the note that follows). As we noted in the previous section, the *vis-viva* controversy and the deep issues of standards of proof and limits of observational technology at the time could be used in a discussion of how science evolves.

CHAPTER 3   GRAVITY

For our K-12 colleagues, we looked at the Next Generation Science Standards to see what topics might benefit from the models and ideas in this chapter. Generally, they fall under MS.Space Systems (www.nextgenscience.org/msess-ss-space-systems) and Earth's Place in the Universe, particularly "MS-ESS1-2 2 develop and use a model to describe the role of gravity in the motions within galaxies and the solar system" (www.nextgenscience.org/ms-ess1-2-earths-place-universe).

Science Fair Project Ideas

There are many ways you can extend some of the ideas in this chapter. Some basic ones would be to apply some of the equations to more detailed models than the simple ones here. For example, if you allow for the fact that the orbits of most of the planets in the solar system are slightly elliptical, what is their orbital velocity at a given point in the orbit? How would the orbit of a comet appear?

A more sophisticated project might involve trying to develop a model of the gravity well of one of the stars that has an eclipsing companion, as Algol does, and then observe for yourself the actual stars eclipsing. You could also use the orbit equation to determine what the period should be if you can look up the orbital parameters. The general principles for Algol are described in a November 19, 2014, *Sky and Telescope* article by Bob King, "When Algol Winks, Will You Wink Back?" www.skyandtelescope.com/astronomy-blogs/behold-algol-star-secret.

Note Be careful applying the *vis-viva* equation, as it only applies in the form given for two-body systems—that is, two bodies in isolation, where you have defined one body to be the origin of a coordinate system and ignoring all other effects. In our solar system, the effects of the other planets are small compared to the Sun, but in star systems with three or more stars (like the Algol system), the orbits of the stars are about their common center of mass, not about one of the stars. Three-body system orbits are not stable indefinitely, and young stars can be in *trapezia*—systems of stars that are not yet in stable orbits around their mutual center of mass. Once we get into thinking about all the mass in the universe and how it interacts, we range into the terrain of Einstein's theory of *general relativity* and a premise known as *Mach's principle*, both of which are way beyond where we want to go here.

You might also think about how you would model escape velocity as well as other types of orbits we have not discussed, like a *hyperbolic* orbit—where something moving very fast just flies close to a gravitating body and then flies into space and never comes back. If you want to explore how spacecraft move from one planet to the other, you can research *Hohmann transfer orbits*. You could also think about what it might be like to be on a planet that had a very elliptical orbit—could life evolve there? What would it be like to live there?

Summary

In this chapter, we learned about gravity and experimented with models of the gravitational potential field around stars and planets. Next, we looked at some ways to model the orbital velocity of planets, moons, and stars in their (elliptical) orbits. These models are mostly special cases of general physics phenomena that require calculus techniques to analyze, so the models need to be taken with a grain of salt. Still, they can be used as starting points for explorations in understanding our own solar system and the dynamics of other systems of distant stars as well.

CHAPTER 4

Airfoils

Flying is so much a part of our lives that it is hard to remember that the ability to fly a heavier-than-air vehicle is barely more than a century old. The Wright brothers succeeded where many others failed largely because of their extremely systematic approach to the problems of flight, and their recognition that the biggest problem to overcome was the control of their craft.

Tip Orville and Wilbur Wright's pragmatic, low-budget approach will be very familiar to today's makers. David McCullough's *The Wright Brothers* (Simon and Schuster, 2015) explores their approach, which could give you some ideas on the type of experiments you might want to do. Or you might just find it inspiring to see how two people with limited resources solved problems that defeated large, well-funded traditional institutions.

A critical part of an airplane's design is its *wings*. The cross section of a wing parallel to the fuselage is called an *airfoil* (or *aerofoil*, if you hail from a place that speaks British English). A modern airplane also usually has parts of the wing that move to help control the airplane. This makes the wing a pretty complex robotic device, not to mention the rest of the plane.

CHAPTER 4 AIRFOILS

The design of a modern airliner wing is a complex undertaking, occupying huge teams of people for years. However, we can get a lot of insight by looking back to a time when things were simpler and a lot more experimentally based than they are now.

This chapter looks at some historic airfoils and gives you an opportunity to 3D print them. The intent here is not to make something flyable as much as to make something that will be a good base for experiments in a classroom or for a science fair project. If you are an aviation history buff, you will be able to create wings of some World War II-era airplanes.

Note The math behind these models is more sophisticated than that behind other models in this book. In one place, a basic calculus idea is used. If you have not yet had calculus, we try to explain what is going on in words. You can use the models without really understanding the math entirely. We will write equations in a mix of conventional algebra (i.e., without using a multiplication symbol between variables) when we are giving a basic physics equation, and pseudocode (using "*" for multiplication) when we are computing a number or transitioning into describing how our 3D printable model works.

CHAPTER 4 AIRFOILS

MODELS USED IN THIS CHAPTER

This chapter uses two different OpenSCAD models. For more on 3D printing and creating and using math models in OpenSCAD, check out Chapter 1. Select 3D printable STL example files are included in the repository as well, and we will describe them in the chapter text. The OpenSCAD models are

- airfoil.scad: This model 3D prints just one wing. It does not support dihedral or sweep without taper (see chapter for definitions of these quantities). The model is shown in the chapter as Listing 4-1.

- airfoilStand.scad: This model prints a pair of wings that need to be glued together, along with an integrated stand that allows it to be used for testing. This model supports more parameters than does the single wing, as described in the chapter. It is shown in the chapter as Listing 4-2.

Some additional materials are also needed to do experiments with the models:

- A box fan (like the one in Figure 4-10) or other means of making a steady breeze
- A postal scale, ideally accurate to 0.1 gram
- Pennies or other small objects to act as balance weights
- A pile of books or other sturdy flat objects to get the model in a smooth part of the fan's airstream
- A hot glue gun or some sturdy tape

CHAPTER 4 AIRFOILS

How Wings Work

In the Wright brothers' day (around 1903) and for some time thereafter, airplane wings were made of fabric and strategic wood stiffeners. Every plane was an experiment, and each wing was laid up and tweaked individually. By the 1930s, engineers started to want to have some standards to work with, and our models will use a template of sorts that was first defined about that time: the NACA airfoils.

Flight Forces: Lift, Gravity, Drag, Thrust

How airplanes fly is complicated. However, we can simplify it down to some basics. Four forces are acting on an airplane. Gravity pulls the plane downward, counteracted by lift pushing the plane upward. Next, there is thrust (the plane's engines) pushing it through the air. The force acting against thrust is drag. Lift and drag are closely related and determined by the geometry of the aircraft and how it interacts with the atmosphere around it.

When a plane is airborne (Figure 4-1) in stable flight, these four forces are in balance, and the airplane moves ahead at a constant speed. If thrust is greater than drag, the aircraft accelerates; if there is enough lift to overcome gravity, the plane can climb.

CHAPTER 4   AIRFOILS

Figure 4-1. *Forces on an aircraft, shown on a model of a World War II British Royal Air Force fighter-bomber, the De Havilland Mosquito—one of the last bombers made of wood. Model courtesy of Stephen Unwin*

The wings have a cross section (the *airfoil*) which is usually not symmetrical. In most airfoils, the top is curved more than the bottom. This makes air on top move faster and be at a lower pressure than air forced across the bottom since air moving faster is less dense than air moving slower. This is not quite the whole story, though.

Wings tend to be rounded at the front (the leading edge) and pointy at the back (the trailing edge). A pointy edge means that air cannot circulate around the wing and get up into the lower pressure area above. Instead, the flow around the wing throws off big vortices (swirling masses of air) that otherwise would have wanted to curl up around the wing. This is called the Kutta Condition. We can calculate (approximately) how much lift a given wing will generate with this equation:

$$Lift = \frac{1}{2} C_L \rho v^2 A$$

105

CHAPTER 4 AIRFOILS

where

- ρ (a Greek letter pronounced "rho") is the density of the air in which the aircraft is traveling, typically given in units of mass per cubic volume.
- v is the velocity at which the wing is passing through a fluid (squared here) in units of distance per unit time.
- A is the planform area of the wing (the area of the wing as seen from above) in units of distance squared.
- C_L is a fudge factor called the lift coefficient that depends on the geometry of the wing and some factors about how the plane is flying. It is usually measured experimentally. You can get some background in the Wikipedia article, "Lift coefficient." It is unitless.

This works out that lift has units of force: Newtons in the international system and pounds-force in Imperial units.

Example: At sea level, if a plane is going 100 miles an hour at takeoff (about 45 m/s), air density at sea level is close to the standard value of 1.23 kg/m³, wing area is around 100 m², and lift coefficient is about 1, then we get about 124 kN (27,000 pounds-force) of lift. The lift coefficient usually varies a lot with the angle the plane's nose is holding relative to the horizontal, called angle of attack (more on this in a later section).

Drag is the resistive force that holds back the airplane as it flies. The equation for drag is the same equation as lift, with a different coefficient (C_D instead of C_L). There is a list of some values of drag coefficient for different shapes in the Wikipedia article, "Drag coefficient." For a plane to fly a long way on a reasonable amount of fuel, it needs to have a big lift coefficient and a small drag coefficient. In practice, they are closely related, and airplane designers have to work to get one as big as possible and push the other down. The ratio of lift to drag is a common performance metric for aircraft.

Chord, Camber, and Thickness

The Wright brothers built on work by others, notably George Cayley (1773–1857), who has been generally credited with defining some of the key features of working wings. If you look at a bird, you see that the wing has a lot of structure and is not just a flat plate tilted to the wind.

Over time, people have developed conventions for ways to describe a wing. There are a lot of subtle issues and complications, but we will just talk about a few of the easiest-to-understand ones in this chapter and give you some suggestions about where to learn more at the end.

Figure 4-2 is a diagram of the parts of an airfoil. The straight-line distance from *leading* (front) edge to *trailing* (back) edge is called the *chord*. In Figure 4-2, the leading edge is on the left. Wings are usually curved from the leading to the trailing edge. The line that describes this curve is called the *camber line*, shown in red in Figure 4-2. By convention distance along the chord is typically the x coordinate. Aircraft typically call the upward direction (toward the sky) z, but since most of our operations here are 2D, our OpenSCAD model uses y for this direction. That way we have a familiar x-y coordinate system to work with.

For simple wings like we are talking about here, camber and thickness are typically specified equations as a function of the x variable running along a chord of length 1. We then just scale the result by the actual length of the chord. In Figure 4-3, the chord is the straight (x axis) line distance from the leading to the trailing edge. The red line right through the middle is the camber line. The short gray lines from the upper to the lower surface, perpendicular to the camber line, represent the thickness of the airfoil at any given point along the camber line.

CHAPTER 4   AIRFOILS

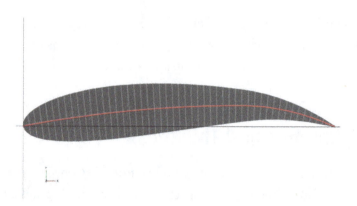

Figure 4-2. *Airfoil (NACA 6715). The red middle line is the camber line*

Note   If you have had some calculus, you can read about the Kutta–Jukowski Theorem to understand a bit better how the phenomenon called circulation creates lift on a wing (in Wikipedia, "Kutta–Jukowski Theorem"). For our purposes here, it is enough to think about a wing throwing off air curving downward behind it to create a lift force. A big, heavy aircraft has so much energy in the swirling air it leaves in its wake that aircraft must be separated by miles to allow these vortices, as they are called, to dissipate. Otherwise, a small plane flying behind a huge one can be flipped over! Search online for videos of "wingtip vortices," and you will never look out an airline window quite the same way again! There is also a good interactive discussion of all these topics in the NASA Glenn Beginner's Guide to Aeronautics noted under "Where to Learn More" at the end of this chapter.

Other Wing Features

Modern wings rarely have straight leading edges, with the same chord and airfoil cross section all the way along the wing. (Some general aviation aircraft are the exception and still have what are called, somewhat dismissively, "Hershey bar" wings.) For wings that do not resemble vending-machine chocolate bars, there are other geometrical features that affect how a wing works.

Sweep

If you look at a modern jet aircraft, the wings sweep back at an angle, rather than having a leading edge that is straight across. This became important once jet aircraft were introduced and planes were going very fast. Jet aircraft now cruise near the speed of sound; right at the speed of sound, a plane creates shock waves that are experienced on the ground as sonic booms and on the plane in a variety of ways. A swept wing allows the plane to fly closer to the speed of sound than it would if it had a straight wing.

Taper

Tapered wings get thinner as they get farther from the fuselage (but maintain the same airfoil cross section, as implemented here). The model in Listing 4-1 requires that swept wings be tapered. You might want to play with the two variables and see how the models change. Taper is the chord at the tip divided by the chord at the root.

CHAPTER 4   AIRFOILS

Dihedral and Twist

If the wings of an airplane bend upward from the root to the tip, we say that the wings "have dihedral" or that the dihedral angle is greater than zero. The angle is defined from a line perpendicular to the chord. If the wings droop down a bit, this is called anhedral.

Another feature of a wing is twist, which is a rotation such that the trailing edge of the aircraft is lower as you go from root to tip, as if a giant hand was holding the wingtip and turning it clockwise while holding the fuselage still.

Dihedral and twist have implications for aircraft stability. Twist is sometimes called *downwash* since the point of a combination of twist and dihedral is to have air "wash" off the wingtip and stick to it better in tough conditions. This gives the pilot better use of control surfaces (more on those in a minute).

Control Surfaces

Modern airplanes, particularly composite wing structures, have elaborate wing geometries incorporating complex interactions of all these features. In addition, parts of the wing can be moved independently. These are collectively called control surfaces (usually also found on the airplane's tail).

Ailerons are hinged parts of the trailing edge (usually farthest from the fuselage or *outboard*) that are used to turn the aircraft. An aileron on one side will be turned down, and the one on the opposite wing will turn up. These devices will sometimes have smaller tabs on them called "*trim tabs*" that are used for fine control.

Flaps are large, hinged tabs that are usually inboard (close to the fuselage). They are tilted down on both wings to give more lift at slow speed, essentially changing the camber line and extending it downward. They also generate a lot more drag. There are many different designs; see the Wikipedia article "Flap (aeronautics)" for a good survey. Hinged tabs that go up into the flow on top of the wing are called *spoilers* (since they ruin the lift) and are used on landing as brakes.

An airplane's tail typically will also have what is essentially a secondary wing attached to it. Hinged surfaces called *elevators* help the plane pitch up and down, and a vertical surface on the tail called a *rudder* will turn right or left, changing its heading. This is different from the banking turn caused by the ailerons, which rolls the aircraft about the fuselage. (In practice, a pilot coordinates the turning forces from ailerons and rudder to maintain good control authority over the airplane.)

Although not strictly speaking a "control surface," another feature you may notice on an airplane are *winglets*. These are vertical surfaces that stick up (or, sometimes up and down) at the end of a wing and manage flow off the wingtip to try and manage drag and control issues. They are passive and not moved around in flight by the pilot but interact with the flow around the aircraft just the same.

Angle of Attack

Airplanes tilt up steeply as they are climbing away from the ground both because they are trying to get away from the ground quickly and also because (to a point) a higher angle relative to the flow generates more lift. The angle of the chord line to the horizontal direction of flight is called *angle of attack*.

Typically, drag also increases with angle of attack, so this is one of the many things to trade off in aircraft design. (We talk about this when we get to the model in Listing 4-2.) Too steep an angle of attack can cause a stall, when the air can no longer smoothly come off the wing and starts to become turbulent. Lift drops off steeply in a stall, and if measures are not taken, the plane can go out of control.

CHAPTER 4   AIRFOILS

NACA Airfoils

After World War I ended, it was clear that some design rules about how to make a good wing were needed to take the next steps in aviation. The Aerodynamic Research Institute of Göttingen, in Germany, did one of the first systematic studies of wings, called from 1923 to 1927.

Meanwhile, in the United States, the *National Advisory Committee on Aeronautics* (*NACA*) was formed in 1915 to coordinate aeronautics research. (In 1958, it was closed and merged into NASA.) In 1932, according to a NASA history chronology (hq.nasa.gov/office/pao/History/Timeline/1930-34.html), a first set of NACA airfoils was published, which are thought to have been inspired to some degree by the German ones. Over the next few years, *wind tunnel* tests (see sidebar) were performed on some of the more promising airfoils. More sets were developed over time and have since been published and put in the public domain.

Fundamentally, the NACA profiles gave an equation for the camber line relative to the chord and another one for how thick the airfoil should be at any point. These equations were experimentally developed, and a lot of the theory about why these were good airfoils came much later.

An airfoil only defines the cross section of the wing. All the other geometrical features (sweep, taper, dihedral, additional control surfaces) are features of the overall wing. The 3D printable wing model we will create has the option for adding some of these features onto a basic World War II–era airfoil design called the NACA Four-Digit Series.

CHAPTER 4 AIRFOILS

WIND TUNNELS

A wind tunnel is, in its most basic form, a box with a fan on one end and an opening on the other that allows an engineer to flow air over a model in a controlled way. A model is held on some sort of support designed to disrupt the airflow as much as possible, called a *sting* (which can just be a stick). The different forces on the wing or airplane are also measured somehow. Sometimes a tunnel will just have a way to mix a little smoke in with the flow so that users can watch what the smoke does when it interacts with the vehicle design.

The wing models we show you how to make in this chapter are designed to allow you to explore how a wing works or just to appreciate some early wing designs as works of engineering art. In the "science fair projects" suggestions at the end of this chapter, we have some links to project sites that talk about how you might build a small student wind tunnel for students to learn about basic aerodynamics.

3D Printable Models

We will work with two different (but similar) OpenSCAD models in this chapter. The first (filename airfoil.scad) produces just one wing using a NACA profile, with some limitations on wing features. This model is in Listing 4-1 later in the chapter.

The other (filename airfoilStand.scad) produces a pair of wings on a sting. The model also produces a stand that can hold the wings at a desired angle of attack and measure lift in an airflow from a fan or other source. This model allows the user to specify the NACA airfoil; the chord and length of the wings; and sweep, taper, dihedral, and twist (if any).

CHAPTER 4   AIRFOILS

This model is in Listing 4-2. Figure 4-3 shows our "air force" with a variety of different parameters we used at an event to teach kids about flight. You can see the fan in the background.

Figure 4-3. *The 3D printed "air force"*

First, we will go into the math behind the models and then examine listings of each. At the end of the chapter, we talk about how to test out the models using a household box fan and a postal scale.

The NACA Four-Digit Series

The first set of NACA airfoil profiles are called the *four-digit profiles* because each profile is defined with four numbers, like NACA 2412 (a classic airfoil that is still used in some general aviation aircraft today). The digits of the airfoil NACA *abcd* number are interpreted as follows:

CHAPTER 4   AIRFOILS

- First digit (*a*): The maximum distance the camber profile goes above the chord (in what we are calling the *y* direction), as a percentage of the chord. You multiply this number by 0.01 times the chord to get the actual distance. If the chord was 1 meter long, for a 2412 airfoil, the maximum camber distance from the chord would be 2 times 0.01 times 1 meter, or 0.02 meters (2 cm).

- Second digit (*b*): The location along the chord, in what we are calling the x direction, starting at the leading edge where the maximum camber occurs as a tenth (not a percentage) of the chord. You multiply this number by 0.1 times the chord to get the location. For our 1 meter chord wing and a 2412 airfoil, this means this location would be 0.4 times 1 meter from the leading edge, or 40 cm from the leading edge.

- Third and fourth digit (cd): The maximum thickness above and below the chord, as a percentage of the chord. There is an equation to plug this number into to get the thickness at any given point along the chord. The maximum thickness in our example would be 12% of 1 meter, or 120 cm on either side of the camber line, and is measured perpendicular to the camber line (see Figure 4-3).

To summarize, our example (NACA 2412, a = 2, b = 4, cd = 12) has a thickness that is 12% of the chord. If the chord is 1 meter long, the maximum distance between the camber line and the chord is 2 cm, and this maximum happens 40% of the chord away from the leading edge of

CHAPTER 4   AIRFOILS

the wing (Figure 4-4). The thickness is distributed symmetrically about the camber line. The maximum thickness point in general will not be at the point of maximum camber. (A brim, used to hold the wing on the 3D printer platform, is visible too.)

Figure 4-4. *NACA 2412 model, as created by the model in Listing 4-1*

In the section later in this chapter, "Thinking About These Models: Learning Like a Maker," we talk about how we went all the way back to a 1935 report to wade through inconsistent naming conventions for these parameters (some people use *mpxx*, some *pmxx*, some other things).

We have relied in our analysis on this report by E.N. Jacobs, K.E. Ward, and R.M. Pinkerton, *The Characteristics of 78 Related Airfoil Sections from Tests in the Variable-Density Wind Tunnel*, (1935) NACA-TR-460, retrieved from NASA Tech Reports Service (NTRS) at https://ntrs.nasa.gov/citations/19930091108. It can also be found just by searching on "NACA-TR-460" on the NTRS or an internet search engine.

CHAPTER 4 AIRFOILS

We did shuffle the equations around a little to make them work better with OpenSCAD. In the upcoming sections, we walk through them from first principles. There are other formulations out there too. We found in many places online that they were simply wrong, so use online sources like people's class notes or Wikipedia with care.

Later other NACA airfoil series were developed that had more complicated profiles. The early ones were intended to be rules of thumb that could be used as guides in the precomputer days. Now wings are sophisticated robotic systems with complicated geometries. We can learn a lot by going back to those simpler times and recreating some of these early experiments.

Math of the NACA Model

The NACA four-digit airfoil model is made up of two curves: a camber line that defines the curve of the centerline of the airfoil and a function that defines how thick the airfoil is on either side of this curve. Confusingly, camber is the distance from the chord to the camber line.

The Camber Line

The camber line is derived from two intersecting parabolas, both of which have a maximum at the point of maximum camber (Figure 4-6). Everything in what follows is computed for a wing with a unit chord (i.e., a chord of length equal to 1 in some units) and has to be scaled proportionally to the actual wing chord. The x axis is along the chord. The equations of the parabolas for the camber line are as follows (where a and b are the first two digits of the NACA profile):

For $x \leq b$:

$$camber(x) = \frac{a}{b^2}\left(2bx - x^2\right)$$

117

For $x \geq b$:

$$camber(x) = \frac{a}{(1-b)^2}\left(1 - 2b + 2bx - x^2\right)$$

Notice that when $x = b$, then either one results in

$$camber(x = b) = a.$$

The point $x = b$ is also where the two parabolas cross, and the camber equation for each of the parabolas gives us a as the value of the camber. Figure 4-5 shows us these two parabolas. The red line is the camber line, the merged halves of the two parabolas. The dashed vertical gray line shows $x = b$, where the camber line switches from one parabola to the other. The maximum is at $x = b$, $y = a$, if the chord is equal to 1.

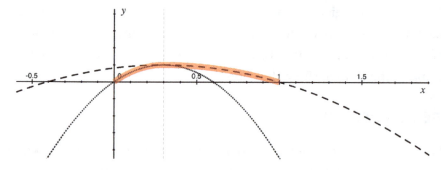

Figure 4-5. *The two parabolas making up the camber line*

The Thickness Equation

The camber line is just an abstract, infinitely thin line. The thickness equation tells us how much the airfoil extends at any point on either side perpendicular to this camber line. The equation for thickness that follows is what you might think of as half the thickness of the wing; it is how far away the wing surface is from the camber line, toward the upper or lower surface.

Some airfoils have different top and bottom thickness profiles, which complicates deriving the camber line. The NACA four-digit ones use the same equation for both, and thus the camber line is a (relatively) simple equation as well. For the four-digit airfoils, the maximum thickness as a percentage of the chord is the last two digits (*cd*) of the NACA airfoil times a function of position along a unit cord, *x*:

$$thickness(x) = \frac{cd}{20}\left(0.2969\sqrt{x} - 0.1260\,x - 0.3516x^2\right)$$

$$+ \frac{cd}{20}\left(0.2843\,x^3 - 0.015x^4\right)$$

However, this is the thickness perpendicular to the camber line and assuming that the chord is 1 unit long. To get it in terms of x and y so we can make an airfoil we can print, we want to figure out what direction is perpendicular to the camber line. An easy way to do this (if you have taken calculus) is to take the derivative (the slope) of a tangent to the camber line.

If this is too much information for you, you can skip the next section and just move on to printing and testing out some wings. If you want to learn more, you might want to check out our book *Make: Calculus* (Make: Community LLC, 2022) for math background. See the "Thinking About These Models: Learning Like a Maker" section for references for these equations, which were in somewhat different form in the 1935 reference.

Calculating the Wing Surface Curve

At this point, we have an equation for camber at any point x along the chord. We also have an equation for the thickness as a function of x. Unfortunately, though, we cannot simply add them. The thickness at any given point needs to be added *perpendicular* to the camber line at that point.

CHAPTER 4   AIRFOILS

To do that, we need to find a tangent line to the camber line at each point. Calculus has a means of finding the tangent line of a curve through a process called "taking its derivative," which is a fancy way of finding a tangent line at any point. When we take the derivative of our camber line (if you have not had calculus yet, just take our word for it), we find that the slope of the camber line as a function of x is

For $x \leq b$:

$$slope(x) = \frac{2a}{b^2}(b-x)$$

For $x \geq b$:

$$slope(x) = \frac{2a}{(1-b)^2}(b-x)$$

These slopes are both zero when $x = b$.

Now, what do we do with these slope lines associated with any point on the camber line? Imagine that we are sliding along the camber line and holding up a stick perpendicular to the slope line we just figured out. The length of the stick is the thickness at that point x, and the tip of the pole would be the wing surface.

To calculate this surface, we define an angle θ (the Greek letter theta) that is the angle the slope line at any point makes with the x axis (Figure 4-6). For each point on the wing's surface, we must raise the point away from the camber line according to our thickness equation in a direction perpendicular to the slope line. We have to do this for each point we calculate along the camber line since its slope is always changing. Note that the blue line will hit the axis to the right of $x = 1$ once we cross the maximum camber point, but the math still applies.

CHAPTER 4 AIRFOILS

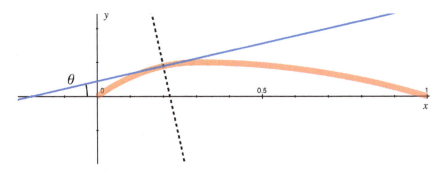

Figure 4-6. *Geometry of finding direction of thickness*

Now for the nitty-gritty of figuring out the details. By using a bit of trigonometry, we can generate the two right triangles shown in Figure 4-7.

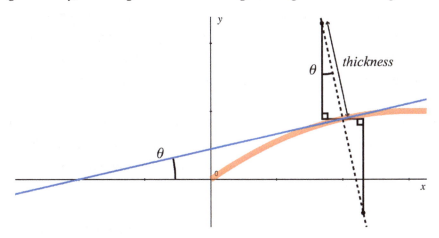

Figure 4-7. *Zooming in on thickness*

The value of *thickness(x)* given by the equation earlier in this section is the vertical side of each triangle. We want the hypotenuse, though, and we want to get the coordinates of the vertex farthest from the camber line for each triangle, for both the upper and lower surface of the wing. Pondering these triangles will give us

121

CHAPTER 4 AIRFOILS

$$\theta = atan(slope(x))$$

$$x_{lower} = x + sin(\theta) * thickness(x)$$

$$y_{lower} = camber(x) - cos(\theta) * thickness(x)$$

$$x_{upper} = x - sin(\theta) * thickness(x)$$

$$y_{upper} = camber(x) + cos(\theta) * thickness(x)$$

Our model then uses all this to compute the wing surface. The more complex model in Listing 4-2, `airfoilStand.scad`, calculates all the points for both the base and tip of the wing, taking other variables like sweep and taper into account. It then uses OpenSCAD's `polyhedron()` module to efficiently stitch these points together with triangles to create the surfaces of the 3D shape.

However, in our OpenSCAD model `airfoil.scad` (Listing 4-1), we do not actually calculate the *x* and *y* coordinates of each point on the upper and lower surface. Instead, we just use the OpenSCAD built-in `rotate()` function to rotate a shape centered on the camber line so that its minimum and maximum points fall on the upper and lower surface lines. Ideally, these shapes would simply be line segments with a length determined by the thickness function. You can see these ideal lines in the gray lines perpendicular to the camber line in Figure 4-2.

However, because OpenSCAD does not have a concept of lines with no area or volume, we simulated these lines with very thin rhomboids (Figure 4-8), which become infinitely thin where they meet the top and bottom surface lines. We created these using a "circle" with four sides with `circle($fn = 4)` and scaling this "square circle" in the x and y directions. `circle($fn = 4)` is a shorter way of producing the same thing we would get by creating a square measuring two units corner to corner and rotating it, which you would do in OpenSCAD with `rotate(45) square(sqrt(2), center = true)`. Note that the rhomboids in Figure 4-8 are thicker and fewer than the ones used to make the wing profile.

CHAPTER 4   AIRFOILS

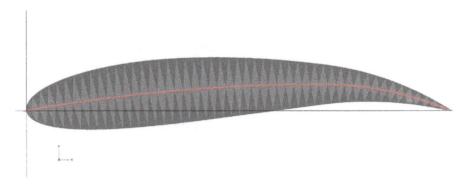

Figure 4-8. The geometry of rhombuses that are bisected by the tangent to the camber line. Airfoil is NACA 6715.

We then use the OpenSCAD convex hull operation hull() to fill in the space between each of these rhomboids and the next to create the profile of our wing. (A convex hull is a shape that incorporates a set of points and creates the minimum convex surface that includes them, in this case our airfoil surface.)

Single Wing OpenSCAD Model

All this results in an airfoil that has a NACA four-digit profile and is a flat wing—not swept back. Listing 4-1 (file airfoil.scad) gives the model for this basic wing, which also has the capability of using sweep and taper.

You can set the following parameters to get the wing you want, for example:

- NACA = 2412;
    - The airfoil desired, 4 digits
- chord = 60;
    - Length of the chord, in mm

123

CHAPTER 4   AIRFOILS

- `length = 120;`
  - Wingspan, in mm (longest distance tip to the root of a single wing, for both models)
- `taper = 1;`
  - How much the wing tapers, ratio tip/root (normally < 1)
- `sweep = 0;`
  - If the wing is swept, sweep angle in degrees; if 0, no sweep

One 3D printable STL file, `airfoil2412.stl`, is included in the repository for this model. It produces a 2412 wing with no sweep or taper.

Listing 4-1. *The Model for a NACA Profile Wing (file* `airfoil.scad`*)*

```
//OpenSCAD model to print out a NACA airfoil
//Formulation of NACA airfoil mathematics based on equations in
//NACA Report 460, "The Characteristics of 78 Related Airfoil
//Sections From Tests in the Variable-Density Wind Tunnel"
//(1935) by E.N. Jacobs, K.E. Ward and R.M. Pinkerton.
//(c) 2016-2024 Rich Cameron
//for the book 3D Printed Science projects, Volume 1
//Licensed under a Creative Commons, Attribution,
//CC-BY 4.0 international license, per
//https://creativecommons.org/licenses/by/4.0/
//Attribute to Rich Cameron, at
//repository github.com/whosawhatsis/3DP-Science-Projects

NACA = 2412; //4-digit NACA airfoil number
chord = 60; //root chord length, mm
length = 120; //wing length, mm
```

```
taper = 1; //ratio of chord at tip over chord at root
sweep = 0; //sweep angle in degrees. 0 = no sweep

assert(
  !(taper == 1 && sweep),
  "ERROR: Sweep without taper is not currently supported!"
);

step = 1/200;

$fs = .5;
$fa = 2;

airfoil();

//Extract the wing parameters from the NACA number
//and return them as an array of [a, b, cd]
function parameters(NACA) = [(NACA - NACA % 1000) / 100000,
  (NACA % 1000 - NACA % 100) / 1000, NACA % 100];

echo("a", parameters(NACA)[0]);
echo("b", parameters(NACA)[1]);
echo("cd", parameters(NACA)[2]);

//Develop the camber line,
function camber(x, p) = (x < p[1]) ?
  p[0] / pow(p[1], 2) * (2 * p[1] * x - pow(x, 2))
:
  p[0] / pow(1 - p[1], 2) *
  (1 - 2 * p[1] + 2 * p[1] * x - pow(x, 2));

//Determine the thickness
function thickness(x, p) = (p[2] / 20) * (
  0.29690 * sqrt(x)
  - 0.12600 * x
```

CHAPTER 4 AIRFOILS

```
    - 0.35160 * pow(x, 2)
    + 0.28430 * pow(x, 3)
    - 0.10150 * pow(x, 4)
);

//Find instantaneous angle of slope of the camber curve, theta,
//so that the thickness component can be computed perpendicular
//to the camber line
function theta(x, p) = atan(
  x < p[1] ?
    p[0] / pow(p[1], 2) * (2 * p[1] - 2 * x)
  :
    p[0] / pow(1 - p[1], 2) * (2 * p[1] - 2 * x)
);

//Draw create the full airfoil from the cross-section by
//extruding the cross-section
module airfoil(
  p = parameters(NACA), chord = chord, length = length,
  taper = taper, sweep = sweep
) {
  translate([length * taper * tan(sweep) + chord / 4, 0, 0])
    linear_extrude(length, center = false, scale = taper)
      translate([-length * taper * tan(sweep) - chord/4, 0, 0])
        airfoil_cross_section(p, chord);
}

//Create the cross-section by hulling a series of rhomboids
//bisected by a tangent to the camber line
//with heights equal to the wing thickness at that point
module airfoil_cross_section(p, chord)
  for(x_ = [0:step:1 - step])
    hull() for(x = [x_, x_ + step])
```

CHAPTER 4 AIRFOILS

```
translate([x * chord, camber(x, p) * chord, 0])
  rotate(theta(x, p))
    if(thickness(x, p))
      scale([chord * step / 10, thickness(x, p) * chord])
        circle($fn = 4);
    else circle(.00001, $fn = 4);
```

Sting and Wings Model

The basic OpenSCAD model in Listing 4-1 will print one wing. If it is not symmetric (if it is swept, for example), you will need to use your 3D printer slicing program (see Chapter 1) to mirror the wing if you want two to make yourself a whole airplane to study.

If you want to study the aerodynamics of a wing and how changing parameters in the model affects lift, Listing 4-2 creates two wings and a test stand (called a *sting*) that will allow it to stand in a wind tunnel or other place where you make observations or measurements. The sting has a ratchet mechanism that allows you to vary the angle of attack in a range that you set.

This model uses the same parameters as does the model in Listing 4-1, plus the variable dihedral, which is a number in degrees. You can also change some parameters for the stand, but most likely you will fare better if you leave those at their defaults unless you have an unusual test situation or want to do something more ambitious than we describe in the next sections.

You will need to glue or tape the two halves together for testing. Low-temperature hot glue works or strategic masking tape or duct tape in a pinch. The latter will alter the wing surface somewhat of course. The model shown in Figure 4-9 used these parameters:

- NACA = 2412;
- chord = 60;

127

CHAPTER 4 AIRFOILS

- length = 120;
  - Note that this is the length of each wing, not the pair
- taper = 0.5;
- sweep = 30;
- dihedral = 0;

and the other parameters at their defaults in Listing 4-2. This is included as the 3D printable STL file airfoilStand2412.stl. A second version with all the same parameters but for NACA airfoil 0015 is included as airfoilStand0015.stl. The 0015 is a symmetrical airfoil; compare how different its lift is with the angle of attack versus the 2412.

Note Some combinations of parameters in these models can make an OpenSCAD render take a very long time in the version of OpenSCAD we used for development (2021.01). Upcoming versions of OpenSCAD may make this faster. Do not be alarmed if it takes as much as an hour to render! Be sure to check the overall look of your model using the Preview function of OpenSCAD (see Chapter 1) before rendering.

Listing 4-2. OpenSCAD Model for a Pair of Wings Plus a Sting (file airfoilStand.scad)

```
//OpenSCAD model to print out a pair of NACA airfoils
//(to make a complete wing)
//Plus a support that can be used for measuring lift
//Formulation of NACA airfoil mathematics based on equations in
//NACA Report 460, "The Characteristics of 78 Related Airfoil
//Sections From Tests in the Variable-Density Wind Tunnel"
```

CHAPTER 4   AIRFOILS

```
//(1935) by E.N. Jacobs, K.E. Ward and R.M. Pinkerton.
//(c) 2016-2024 Rich Cameron
//for the book 3D Printed Science projects, Volume 1
//Licensed under a Creative Commons, Attribution,
//CC-BY 4.0 international license, per
//https://creativecommons.org/licenses/by/4.0/
//Attribute to Rich Cameron, at
//repository github.com/whosawhatsis/3DP-Science-Projects

NACA = 2412; //4-digit NACA airfoil number
chord = 60; //root chord length, mm
length = 120; //wing length, mm

taper = 1; //ratio of chord at tip over chord at root
sweep = 0; //sweep angle in degrees. 0 = no sweep
dihedral = 0; //dihedral angle in degrees. 0 = no dihedral

sting_size = 20; //sting cross section in mm
sting_length = 100; //sting vertical bar length, in mm

sting_angle = [0:5:25]; //range of angle of attack possible

tolerance = .3;
nest = true; //Attempt to nest parts for printing
//Some values may result in nested parts overlapping

steps = 200;
step = 1/steps;

$fs = .5;
$fa = 2;

airfoil_with_sting();
%mirror([0, 0, 1]) airfoil_with_sting();
%translate([
```

CHAPTER 4   AIRFOILS

```
      chord * 1.5 + sting_size/4,
      -sting_length,
      -sting_size/2
    ]) base();

    if(nest) {
      translate([
        chord * 1.5 - sting_size / 2 - 5,
        -65,
        0
      ]) rotate(-65) base();
      translate([
        chord * 2.5 - max(chord, 40) - sting_size / 2,
        sting_size / 2,
        0
      ]) mirror([1, 0, 0]) airfoil_with_sting();
    } else {
      translate([0, 50, 0]) base();
      translate([-5, 0, 0]) mirror([1, 0, 0]) airfoil_with_sting();
    }

//Extract the wing parameters from the NACA number
//and return them as an array of [a, b, cd]
function parameters(NACA) = [(NACA - NACA % 1000) / 100000,
    (NACA % 1000 - NACA % 100) / 1000, NACA % 100];

echo("a", parameters(NACA)[0]);
echo("b", parameters(NACA)[1]);
echo("cd", parameters(NACA)[2]);

//Develop the camber line,
function camber(x, p) = (x < p[1]) ?
    p[0] / pow(p[1], 2) * (2 * p[1] * x - pow(x, 2))
    :
```

```
    p[0] / pow(1 - p[1], 2) *
    (1 - 2 * p[1] + 2 * p[1] * x - pow(x, 2));

//Determine the thickness
function thickness(x, p) = (p[2] / 20) * (
  0.29690 * sqrt(x)
  - 0.12600 * x
  - 0.35160 * pow(x, 2)
  + 0.28430 * pow(x, 3)
  - 0.10150 * pow(x, 4)
);

//Find instantaneous angle of slope of the camber curve, theta,
//so that the thickness component can be computed perpendicular
//to the camber line
function theta(x, p) = atan(
  x < p[1] ?
    p[0] / pow(p[1], 2) * (2 * p[1] - 2 * x)
  :
    p[0] / pow(1 - p[1], 2) * (2 * p[1] - 2 * x)
);

//Utility function for converting 2D points to 3D
function z(points, z = 0) = [for(p = points) [p[0], p[1], z]];

//Utility function for adding to all points in an array
function add(v, plus) = [for(i = v) i + plus];

//Calculates surface points to be stitched together later
function airfoil_points(p, chord) = concat(
  //top
  [for(x = [0:step:1 - step]) chord * [
    x - thickness(x, p) * sin(theta(x, p)),
    camber(x, p) + thickness(x, p) * cos(theta(x, p))
```

CHAPTER 4 AIRFOILS

```
    ]],
    //bottom
    [for(x = [1:-step:step]) chord * [
      x + thickness(x, p) * sin(theta(x, p)),
      camber(x, p) - thickness(x, p) * cos(theta(x, p))
    ]]
  );
module airfoil_cross_section(p, chord)
  polygon(airfoil_points(p, chord));

//Build the wing by stitching together the points of the root
//and tip cross-sections. This is the fastest method to
//compute, and allows all combinations of sweep and taper.
module airfoil(
  p = parameters(NACA), chord = chord, length = length,
  taper = taper, sweep = sweep, dihedral = dihedral
) {
  polyhedron(
    concat(
      z(airfoil_points(p, chord)),
      z(
        add(
          add(
            airfoil_points(p, chord),
            [-chord / 4, 0]
          ) * taper,
          [
            chord / 4 + length * tan(sweep),
            length * tan(dihedral)]
        ), length
      )
```

```
      ), concat(
        [for(i = [1:steps - 1]) [i, steps * 2 - i, i + 1]],
        [for(i = [1:steps - 1]) [i,
          (steps * 2 - i + 1) % (steps * 2), steps * 2 - i]],
        [for(i = [1:steps - 1]) add([i, i + 1,
          steps * 2 - i], steps * 2)],
        [for(i = [1:steps - 1]) add([i, steps * 2 - i,
          (steps * 2 - i + 1) % (steps * 2)], steps * 2)],
        [for(i = [0:steps * 2 - 1]) [i, (i + 1) % (steps * 2),
          (i + 1) % (steps * 2) + steps * 2]],
        [for(i = [0:steps * 2 - 1]) [i,
          (i + 1) % (steps * 2) + steps * 2, i + steps * 2]]
      )
  );
}
*translate([chord / 4, 0, 0]) rotate([0, sweep, 0]) %cube(100);

module airfoil_with_sting() translate([0, 0, 0]) union() {
  airfoil();
  hull() {
    translate([chord * 0.3, chord * parameters(NACA)[0], 0])
      intersection() {
        sphere(r = chord * thickness(0.3,
          parameters(NACA)));
          translate([0, 0, chord/2])
            cube(chord, center = true);
      }
    translate([
      chord * 1.5 - sting_size/4,
      chord * parameters(NACA)[0],
      0
    ])
```

CHAPTER 4   AIRFOILS

```
      intersection() {
        rotate([0, -90, 0])
          rotate_extrude() rotate(-90) intersection() {
            airfoil_cross_section(
              parameters(0040),
              sting_size
            );
            square(sting_size);
          }
        translate([0, 0, sting_size])
          cube(sting_size * 2, center = true);
    }
  }
  difference() {
    translate([
      chord * 1.5 - sting_size/4,
      chord * parameters(NACA)[0],
      0
    ])
      rotate([90, 0, 0])
        linear_extrude(sting_length) intersection() {
          airfoil_cross_section(parameters(0040), sting_size);
          square(sting_size);
        }
    translate([chord * 1.5 + sting_size/4, -sting_length, -1])
      cylinder(r = sting_size/2 - 3, h = sting_size);
  }
  translate([chord * 1.5 + sting_size/4, -sting_length, 0])
    intersection() {
      rotate([-90, 0, 0]) rotate_extrude(convexity = 10)
        hull() {
```

```
      difference() {
        circle(r = sting_size / 2 + 1);
        translate([0, -50, 0]) square(100);
      }
      translate([-50, 0, 0]) circle(r = 1);
    }
    linear_extrude(sting_size / 4, convexity = 10)
      difference() {
        hull() {
          circle(r = sting_size / 2);
          translate([-50, 0, 0]) circle(r = 1);
        }
        circle(r = sting_size/2 - 4, $fn = 8);
      }
    }
  }
}
module base() difference() {
  union() {
    difference() {
      linear_extrude(sting_size, convexity = 10) difference() {
        intersection() {
          circle(50 + 5);
          translate([-100 + sting_size, 100 - sting_size, 0])
            square(200, center = true);
          rotate(-max([for(i = sting_angle) i]))
            translate([-100 + sting_size, -100 + 5, 0])
              square(200, center = true);
        }
        circle(50 - 2);
        for(i = sting_angle) rotate(-i) hull() {
          circle(r = sting_size / 2 + 3);
```

```
          translate([-50, 0, 0]) circle(r = 1 + tolerance);
        }
      }
      for(i = sting_angle) rotate(-i) hull() for(j = [1, -1])
        translate([0, 0, sting_size + 5 * j - 3])
          linear_extrude(1) offset(1.5 * j)
            hull() {
              circle(r = sting_size / 2 + 3);
              translate([-50, 0, 0]) circle(r = 1.1);
            }
    }
    linear_extrude(sting_size / 4) difference() {
      intersection() {
        circle(50 + 5);
        translate([-100 + sting_size, 100 - sting_size, 0])
          square(200, center = true);
        rotate(-max([for(i = sting_angle) i]))
          translate([-100 + sting_size, -100 + 5, 0])
            square(200, center = true);
      }
    }
    difference() {
      intersection() {
        linear_extrude(73) difference() {
          intersection() {
            circle(50 + 5);
            translate([-100 + sting_size, 7 - sting_size, 0])
              square([200, 14], center = true);
            rotate(-max([for(i = sting_angle) i]))
              translate([-100 + sting_size, -100 + 5, 0])
                square(200, center = true);
          }
```

```
      }
      translate([-100, 0, 80]) rotate([-130, 0, 0])
        cube(200);
    }
    intersection() {
      linear_extrude(100) difference() {
        intersection() {
          circle(50 + 5 - 3);
          translate([
            -100 + sting_size,
            7 - sting_size + 3,
            0
          ]) square([200, 14], center = true);
          rotate(-max([for(i = sting_angle) i]))
            translate([-100 + sting_size, -100 + 5, 0])
              square([200 - 6, 200], center = true);
        }
      }
      translate([-100, 4, 80]) rotate([-130, 0, 0])
        cube(200);
    }
  }
  difference() {
    hull() {
      cylinder(
        h = sting_size - 2,
        r = (sting_size/2 - 4) * cos(180 / 8) - tolerance
      );
      cylinder(
        h = sting_size,
        r = (sting_size/2 - 4) * cos(180 / 8) - tolerance - 1
      );
```

CHAPTER 4   AIRFOILS

```
      }
    }
  }
  translate([0, 0, 1]) linear_extrude(sting_size - 2)
    for(a = [0:60:179]) rotate(a) square([
      ((sting_size/2 - 4) * cos(180 / 8) - tolerance) * 2 - 3,
      .1
    ], center = true);
}
```

Measuring Lift

The model in Listing 4-2 creates a pair of wings, a sting, and a stand that you can weigh down (with coins, for instance). You also need to attach the two wings to each other, say, with masking tape or by gluing. The base will let you set an angle of attack range. In the example, it goes from 0 to 25 degrees in 5-degree increments. Figure 4-9 shows the wing halves glued together and the stand, and Figure 4-10 shows the model fully assembled.

Figure 4-9. *Model with wings glued together*

CHAPTER 4   AIRFOILS

Figure 4-10. Model fully assembled

Then, if you put the whole thing on a sensitive scale (like an electronic postal scale or a scale you have in your school lab), you will get a setup that looks like the one in Figure 4-11 (shown here with a different wing). Figure 4-12 is a close-up of the rachet for aligning the wing at a desired angle of attack.

Caution   The sting model is intended to allow for some basic classroom experiments in changing the angle of attack and measuring the resulting lift. It is not very aerodynamic and introduces a lot of drag, and we have not talked about measuring drag. If you are interested in measuring anything other than lift, we suggest going back to the OpenSCAD model and adding your own attachments for more sophisticated instruments.

CHAPTER 4 AIRFOILS

Figure 4-11. *Placing a NACA 6721 swept, tapered wing and sting in front of a fan on a postal scale*

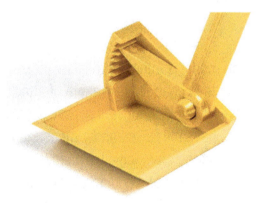

Figure 4-12. *Close-up of the angle of attack ratchet*

If you zero out the scale when the fan is off, you should get a negative number for the weight of the sting plus the airfoil when you turn the fan on. That number is the lift (with a fair number of sources of error). You can use the ratchet shown in Figure 4-12 to vary the angle of attack, and of course, you can print different wings.

You will need to find some way of measuring how fast the flow coming out of the fan is and its direction—you can make, buy, or borrow a cup anemometer as a simple instrument (a search will reveal both plans to make and places to purchase one). As an admittedly dubious fallback, you can take a small piece of string and hold it in front of the wing to see if you are in an area of the fan outflow that has smooth airflow. We guesstimated the flow out of our fan by dropping a small scrap of tissue, seeing how long it took to go a few meters, and averaging the results of a few trials. This is too error-prone to use closer than a factor of two or so since the flow mixes very quickly with the ambient still air. But it is a good start.

In our experiment with a 2412 wing with a chord of 60 mm, a taper of 0.5, and a wing length (each wing) of 80 mm, an angle of attack of 0 degrees gave us enough lift to raise about 10 grams; of 15 degrees, about 13; and of 25 degrees, about 8 grams. The 1935 paper seems to show that a clean 2412 wing should have a lift coefficient of about 0.2 at 0 degrees angle of attack and about 1.3 at 15 degrees, and at around 25 degrees or so, it should stall and drop off precipitously. Given the many variables (like the effects of the test stand and the placement in the fan's not-particularly-even flow), this is reasonable. We encourage you to work out ways to make this far more accurate!

Household fans often have dead spots. You may need to play around a bit to get good results—staying a couple feet away from the fan helps, but not too far, since the flow will be too diffuse. To do this right, you need to create a wind tunnel, the device that controls air around a test airplane. See the "Building a Student Wind Tunnel" resources at the end of the chapter.

CHAPTER 4   AIRFOILS

If you want to compare your measured lift against a theoretical value, earlier in the chapter, we saw that lift can be approximated as

$$Lift = \frac{1}{2} C_L \rho v^2 A$$

Reasonable numbers for the variables are as follows:

- ρ, the density of air, is about 1.23 kg per cubic meter at sea level (adjust if you are in Denver or someplace else significantly above sea level).
- C_L, the lift coefficient, will vary with angle of attack but will be around 1 or 2 (see the 1935 NACA Tech Report).

To get the planform area A of a tapered wing, you must figure out the area of a trapezoid with the sides being the root chord and tip chord. To do that, average the two chords and multiply by the wingspan. In the case of our example, we would get

Area = 0.5 * (tip chord + root chord) * wingspan

In our example here,

Area = 0.5 * (60 mm + 30 mm) * 80 mm * 2 wings

= 0.0072 square meters.

If the velocity v of the airflow is about 9 miles per hour (about 4 meters per second), then the lift should be around 0.5 * 1 * 1.23 * 4 * 4 * 0.0072 = 0.0709 Newtons, or the ability to lift about 7 grams if you prefer to think that way, for a drag coefficient equal to 1.

Tip For more ideas on simple ways to measure lift and find other planform areas, see the slides "Wind tunnel experiments for Grades 8–12," available at www.grc.nasa.gov/www/k-12/airplane/topics.htm.

Note These models are not meant to be flying models. They will barely lift their own weight at reasonable speeds. They are meant to be sturdy enough to stand up to some experimentation and analysis, rather than being as light as possible. Your 3D printer software will give you a weight estimate for the wing (which in the case of the Listing 4-2 models will include the sting and the base).

Printing Suggestions

These models are meant to be as easy as possible to print, given that the geometry is complicated. Print them with the chord side down, as the models were designed. You should not need to use support, but you probably want to use a brim or raft so that the model sticks to the platform.

To fit the STL output by Listing 4-2 on your printer, you may need to move the pieces around to fit your platform or print it in more than one print. You will want to make sure that you have printer settings that allow the brim or raft to be removed cleanly, though, because any remaining pieces will interfere with the assembly of the sting mechanism. Figure 4-13 shows how the model appeared on the 3D printer platform.

CHAPTER 4   AIRFOILS

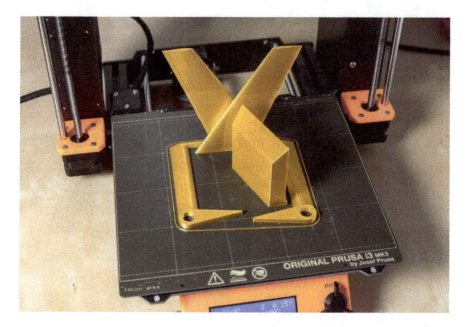

Figure 4-13. The layout of the models on a 3D printer

Caution   Do not scale these models in your 3D printing program; use the scaling parameters in the models in OpenSCAD and generate a new STL. There are issues of tolerances and some complex interactions that may break if you just scale without taking account of these issues.

Classic Airplanes Using NACA Airfoils

The NACA models were developed a little before World War II. All the airplanes flying in those days were what we would now consider to be low speed. The NACA airfoils were used heavily in World War II airplanes on all sides of the conflict. The Incomplete Guide to Airfoil

CHAPTER 4 AIRFOILS

Usage, at http://m-selig.ae.illinois.edu/ads/aircraft.html, has many examples. The same group's home page at http://m-selig.ae.illinois.edu/ads.html has links to other aircraft modeling tools and sites.

You will see in the *Incomplete Guide* that many airplanes used one airfoil design at the root of the wing (where it connects to the fuselage) and another at the tip. For example, according to the *Incomplete Guide*, the workhorse early prop-driven airliner DC-3 used NACA 2215 at the root and 4412 at the tip. Many Cessnas have used the common 2412 airfoil.

If you are a fan of British aviation history, the *Incomplete Guide* notes that the World War II Supermarine Spitfire used two NACA profiles in its wings (2213 at its root, 2209.4 at its tip, where the decimal point means 09.4% maximum thickness). The Spitfire was the classic Royal Air Force (RAF) interceptor during the Battle of Britain. Figure 4-14 is a kit created in the 1960s by the child of a WWII RAF officer.

We have not created the capability to replicate the exact details of any given actual wing (the model requires that you have one NACA profile for the whole wing). Rather, our intent is that you can pretend to be the designer of a historic vehicle and play around with changing its basic profile. In short, how might you have designed an airplane during World War II? Or for a low-speed application today? Or to understand why different birds have the wing shapes they do?

145

CHAPTER 4   AIRFOILS

Figure 4-14. WWII Supermarine Spitfire model (courtesy of Stephen Unwin)

THINKING ABOUT THESE MODELS: LEARNING LIKE A MAKER

When we started this chapter, we thought we would be done in a couple of hours. We knew there were equations for the NACA airfoils, and we thought we would use techniques like those we used in Chapter 1 to create the top and bottom as 3D surfaces.

No such luck.

The NACA airfoils are interesting in that they were created to make hand calculations (in the 1930s) as simple as possible and were purely empirically based. They were extensively cited, but inconsistently. Over time, authors seem to have lost some of the assumptions underlying the original model or just appeared to make transcription errors. Some authors used the letter *p* for the first digit and *m* for the second; somewhere in the 1970s, this

CHAPTER 4 AIRFOILS

seemed to reverse. (To avoid getting a lot of letters from people who found references one way or the other, we changed here to an abcd convention.) Authors inconsistently used either the ratio of the thickness to the chord or the dimensioned number.

We decided to go all the way back to the oldest possible description of the NACA series and to look at the physics there. Fortunately, all these old reports are available for free from the NASA Tech Reports Service, as we referenced earlier in the chapter. We started digging and discovered that the report everyone seemed to cite as the first place the airfoils were described in detail was NACA-TR-460, published in 1935 by E.N. Jacobs, K.E. Ward, and R.M. Pinkerton. This report, *The Characteristics of 78 Related Airfoil Sections from Tests in the Variable-Density Wind Tunnel*, is an amazing period piece. Two of the advisors listed in the front matter were Charles Lindbergh and Orville Wright! Pages 4 and 5 of the report have very clear drawings that we used as the basis of the OpenSCAD model.

If you look in the back of the 1935 report (with its many hand-drawn graphs), you can see wind tunnel data that might be a good place for you to begin designing a classroom demonstration or a great science fair project.

Where to Learn More

There is a tremendous amount of information out there about airfoils; the challenge for this chapter has been to curate a set of interesting explorations that would be accessible to a wide range of student levels. In this section, we suggest some sensible bigger projects for you to take on if just informally playing with these airfoils is not enough. Another good general site on the topic, in addition to the ones given so far in this chapter, is http://airfoiltools.com. This site has a variety of calculators and plotting tools.

CHAPTER 4 AIRFOILS

NASA Glenn Research Center also has an exhaustive site with many good ideas. The main index is at www.grc.nasa.gov/WWW/K-12/airplane/bgt.html, and many presentations available for teachers are indexed at www.grc.nasa.gov/WWW/K-12/ airplane/topics.htm.

You can take projects about airfoils in several directions. As a first thought, you can look at it as the history of aviation and the study of the development of aviation from 1903 through the 1930s and beyond. Or you can treat this as a physics exercise to see how changing some of these basic parameters like camber and thickness changes the lift and drag of the wing.

Visualizing Flow

To learn more about ways to visualize what the air flowing over wings is doing, try searching on "flow visualization." There is also a classic book you might hunt down: *An Album of Fluid Motion*, by Milton Van Dyke (Parabolic Press, 1982), which is just an entire book of pictures of flow in various test conditions.

Building a Student Wind Tunnel

To test out the wings in a quantitative way, you will need to make yourself some sort of wind tunnel. We have not tried this ourselves, but with that caveat, we will note that this project seems plausible, although probably best suited for a group project: sciencebuddies.org/science-fair-projects/references/how-to-build-a-wind-tunnel. To build a tunnel purely to visualize the flow around the wing (versus trying to measure lift and drag), this Instructable seems simpler: www.instructables.com/id/DIY-Wind-Tunnel-20-Project-Paperclip/.

CHAPTER 4   AIRFOILS

Scaling a Model

If you are going to seriously try to model a real system (like a bird or an airplane) with a wind tunnel model that is a lot smaller than the original, you will need to match the Reynolds number (R_e) of your model and the real system to get accurate results. The Reynolds number is a ratio of the kinetic energy in a system to the effects of viscosity. Viscosity is how "sticky" a fluid is—maple syrup is viscous, but air less so. (Engineers do not draw distinctions between liquids and gasses here—it is all "fluids" to us.)

High R_e flow could be found in a fast-flowing water stream or a jet flying through the air. Low R_e is exemplified by molasses coming out of a jar or very small objects not moving very fast. The equation for the Reynolds number is usually written as

$$R_e = \frac{\rho v^2 L}{\mu}$$

where ρ is still the density of the air (or fluid) around the wing, v is the velocity, L is a characteristic length of the thing you are testing (like the chord or the wingspan), and μ is the dynamic viscosity of the air or other fluid (a number you would typically look up for the conditions you are anticipating).

If you are modeling something big with a 1/10th scale model, to have the Reynolds number of your model be the same as the real thing, you need to run your test with about 3 times (square root of 10) higher velocity or ten times denser air or less viscous flow (or a combination that works out to that). For air at the surface of the Earth at 15 degrees Centigrade, the density and viscosity are, respectively, $\rho = 1.23$ kg/m³ and $\mu = 1.78 \times 10^{-5}$ kg/m-s; for water, $\rho = 999$ kg/m³; and $\mu = 115.4 \times 10^{-5}$ kg/m-s (from A.M. Kuethe and C-Y Chow, *Foundations of Aerodynamics 3rd Edition*, Wiley, 1976).

Using flowing water as a test fluid is thus the same as air at the same velocity around an object about 12 times bigger. Water tunnels have been developed to exploit this, but obviously, this is challenging and messy.

Teacher Tips

The science of flight is usually not encountered in depth until undergraduate-level science courses, since to get very far into the analysis, you really need to have some calculus. That said, the empirical study of the forces on a wing might fit well under these NGSS standards: MS-PS-2, Motion and Stability: Forces and Actions, nextgenscience.org/dci-arrangement/ms-ps2-motion-and-stability-forces-and-interactions, and MS-PS-3, Energy standards, nextgenscience.org/pe/4-ps3-1-energy.

That said, there are many topics that could be covered purely to build intuition. We have suggested many possible experiments in this chapter; some might lend themselves to labs at almost any level. Working to refine how accurate the model is will be a good long-term project for any number of students.

Science Fair Project Ideas

A wind tunnel, even the small ones described in the "Where to Learn More" section, is probably physically too big to bring to most science fairs. However, if a school or class builds one, testing a series of models to find their lift and drag might be an interesting project. You might also see which airfoils could approximate a bird's wing and try to think about how to model how well a bird "should" fly.

But if all that sounds too complicated, you can always create a few airfoils and just test them qualitatively in front of a box fan. Consider ways to make simple ways to measure lift and drag using weights to balance

CHAPTER 4   AIRFOILS

the forces. Or perhaps create some wing cross sections to see which ones might match various bird wings. Or, for that matter, you can just drag them through a bathtub and see if you can feel any difference.

Summary

In this chapter, we talked about airfoils and offered a simple model that was used in the 1930s to learn more about how wings work. We discovered that the mathematics behind flight are complex but that there are some simplified models, like the NACA airfoils, that can help build intuition (and that are still in use today). We looked at how we might use those models today to learn about flight and perhaps to help understand birds or other flying creatures.

CHAPTER 5

Simple Machines

Simple machines are devices that change the amount, or magnitude, of force or the direction of a force being exerted on something. The standard list of simple machines (at least as far as school science standards define them) is the pulley, screw, wheel and axle, inclined plane, wedge, and lever. Most of these are very old—the wheel's origin is lost in antiquity, and Archimedes knew about the lever, screw, and pulley about 2,300 years ago.

The physics behind simple machines, though, was not clarified in some cases till the late 1700s. There is a good history of these technologies on the Wikipedia page, "Simple machine." The force-lessening effect that these machines exploit is called *mechanical advantage*. Often it is manifested by moving an object over a distance more slowly but with less force than one would without the intervention of a machine—for example, pushing something up a ramp (inclined plane) versus lifting it straight up.

MODELS USED IN THIS CHAPTER

This chapter uses our different OpenSCAD models. For more on 3D printing and creating and using math models in OpenSCAD, check out Chapter 1. Select 3D printable STL example files are included in the repository as well. The OpenSCAD models are

CHAPTER 5   SIMPLE MACHINES

- `wedgePlane.scad`: Creates a wedge or inclined plane, depending on parameters
- `lever.scad`: Creates various types of levers
- `archimedesScrew.scad`: Creates a model of an Archimedes screw)
- `pulley.scad`: Creates a model of pulleys
- `skewerWheel.scad`: Creates wheels that fit on a skewer
- `chassis.scad`: Creates a toy car chassis

The models `vise.scad` and `pulleysWheels.scad` from the first edition of this book have been retired.

To make all the models in this chapter, you will also need a few of each of the following:

- Wooden (or bamboo) disposable cooking skewers (approximate diameter 3 mm or 1/8") to use as axles in pulley and wheel models
- Alternatively, radial bearings with compatible screws, as will be described in the chapter
- If you prefer to use craft materials for the car chassis instead of creating one with chassis.scad (wheel and axle section):
    - Plastic drinking straws of slightly larger diameter than the skewers
    - Craft (popsicle) sticks
    - Masking tape
- Heavy string or parachute cord
- Something small and heavy to use as a weight (like a few washers)
- A carabiner, or just a bent paperclip, to hook the weight to pulleys

The Machines

Most of the things we normally think of as machines are combinations of basic simple machines. Using more than one simple machine creates a *compound machine*. For example, a wheelbarrow is a combination of a lever (the handle) and two wheels on an axle (at least, for the version we show in a later figure).

Typically, simple machine discussions focus on unpowered machines—ones that rely on human (or animal) power that is then converted into mechanical energy of some sort, with or without amplification, based on the geometries of the machine.

In this chapter, we have created basic versions of the six simple machines in OpenSCAD with the intent that you can 3D print them individually to use as demonstrations. The OpenSCAD model for each is designed so that you can vary critical dimensions to build intuition about how the effects of each machine change as a result.

Each machine may have some caveats about scaling, which we talk about as we introduce them. You will also want to experiment with combining simple machines into compound ones. We give one example of a combination after we introduce all six machines. *Friction* and *flexing* of the machine lower their efficiency to a greater or lesser degree, and you should bear that in mind if your experiments do not achieve the mechanical advantage that your calculations might indicate.

We developed OpenSCAD models for each of the six simple machines, and in the next few sections, we go over the process of 3D printing them one at a time. You can also use them as building blocks to create more complex devices. But be sure to look at the notes about combining the machines later in this chapter for some things to think about, particularly if you want to scale them or to print several of them at once in varying orientations. Chapter 9 shows you how to combine some of these in more complex mechanisms.

CHAPTER 5　SIMPLE MACHINES

Note We give some examples of each machine and a bit about its history as we go. If you are interested in the history of basic inventions like these in general, the book *Cathedral, Forge, and Waterwheel* by Frances and Joseph Gies (HarperCollins, 1994) has in-depth discussions of the paths that some machines took. The ancient Greeks in about the third century BC thought about five of the simple machines (they did not think of the inclined plane as a "machine"). The Egyptians used the inclined plane and lever (but not the wheel and axle) about 4,600 years ago to create the Great Pyramid of Giza. There are Wikipedia articles about each of the simple machines that list more history as well.

Inclined Plane and Wedge

Inclined planes and *wedges* are similar, and the model we created makes one of each at the same time. There are check boxes for each in OpenSCAD's Customizer so that you can choose to only print one of the two. Here, we talk about each machine first and then go over its 3D printable models.

Inclined Plane

The inclined plane creates mechanical advantage by allowing you to move something upward at an angle so that you can exert a lower force than you would if you had to raise it vertically or more steeply. Alternatively, you can convert gravitational potential energy into kinetic energy, as in downhill skiing. The cost may be that you need to move the object over a longer distance. A ramp is an inclined plane, as is a flight of stairs. Think about how much easier it is to walk up a trail with switchbacks versus going straight up a mountain.

CHAPTER 5   SIMPLE MACHINES

The mechanical advantage of an inclined plane is the horizontal distance it spans divided by its maximum height. The longer and shallower the ramp, in other words, the greater the advantage. A ramp that is ten feet long (diagonally) and rises two feet has a mechanical advantage of five to one, ignoring friction and other imperfections. The cost is that the user needs to traverse a longer distance to go up the same vertical rise. Wikipedia's article, "Inclined plane," has some very lucid explanations.

Wedge

A wedge is closely related to the inclined plane. A wedge can either force apart two objects or can hold something in place. Axes and knives are wedges, as is the end of a crowbar (the rest of it being a lever).

Frictional force is important with a wedge, and although the mechanical advantage here is also length divided by maximum width, how well the wedge stays put or forces in without bending matters a lot. Figure 5-1 shows the two "machines." Both have a 30-degree angle, but the inclined plane is a right triangle, and the wedge is an isosceles triangle to make it easier to push into place between things. The models will by default print in the orientation in Figure 5-1.

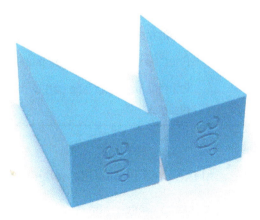

Figure 5-1. *Inclined plane (L) and wedge (R), printed on their sides*

157

CHAPTER 5   SIMPLE MACHINES

Wedge and Inclined Plane Models

The model in Listing 5-1 generates both an inclined plane and a wedge (described in the next section) since one is an adaptation of the other. The model has five parameters you can change, listed in Table 5-1. We recommend that you do not make the wedge or plane a lot smaller than about half the default size shown here, since then the numbers may not print well. You can make both the wedge and plane or just one or the other by setting the values of wedge and plane appropriately.

You do not really need a 3D print to have an inclined plane. Any tilted surface will do, like putting something under the legs on one side of a table or raising one end of a book by resting it on something.

Table 5-1. *Wedge and Inclined Plane Variables*

Variable	Default Value and Units	Meaning
length	100 mm	Length of the plane's base and of the two equal sides of the wedge
angle	30 degrees	Angle of the plane and wedge (needs to be less than 90 degrees)
width	50 mm	Width measured perpendicular to the triangle
wedge	true	Print the wedge part of the model (false to disable)
plane	true	Print the inclined plane part of the model (false to disable)

Listing 5-1. Wedge and Inclined Plane (file wedgePlane.scad)

```
//OpenSCAD model to create a wedge and inclined plane
//File wedgePlane.scad
//(c) 2016-2024 Rich Cameron
```

CHAPTER 5 ■ SIMPLE MACHINES

```
//for the book 3D Printed Science projects, Volume 1
//Licensed under a Creative Commons, Attribution,
//CC-BY 4.0 international license, per
//https://creativecommons.org/licenses/by/4.0/
//Attribute to Rich Cameron, at
//repository github.com/whosawhatsis/3DP-Science-Projects

//length of the plane's base, and of the wedge's  equal sides
length = 100;
//angle of the plane or wedge, in degrees
angle = 30;
//width measured perpendicular to the triangle
width = 50;
wedge = true;
plane = true;

if(wedge) wedge(length, width, angle);
if(plane) translate([0, -length * tan(angle) - 5, 0])
    inclined_plane(length, width, angle);

module wedge(
    length = 100,
    width = 50,
    angle = 30,
        isosceles = true
) difference() {
    linear_extrude(width, convexity = 5) intersection() {
        square([length, length * tan(angle)]);
        rotate(angle - 90)
            square([length * sin(angle), length / cos(angle)]);
        if( isosceles) rotate(angle / 2 - 90)
            translate([-length * sin(angle / 2), 0, 0])
                square(length * [sin(angle / 2) * 2, cos(angle / 2)]);
```

CHAPTER 5   SIMPLE MACHINES

```
    }
    if( isosceles) translate([length, 0, width / 2])
       rotate([0, 90, angle / 2])
         translate([0, length * sin(angle / 2), 0])
            linear_extrude(height = 1, center = true)
              text(
                 str(angle, "°"),
                 size = min(width / 3,
                 length * sin(angle / 2)),
                 halign = "center",
                 valign = "center"
              );
    else translate([length, length * tan(angle) / 2, width / 2])
       rotate([0, 90, 0])
         linear_extrude(height = 1, center = true)
            text(
              str(angle, "°"),
              size = min(width / 3,
              length * tan(angle) / 2),
              halign = "center",
              valign = "center"
            );
}

module inclined_plane(length = 100, width = 50, angle = 30)
    wedge(
       length = length,
       width = width,
       angle = angle,
         isosceles = false
    );
```

CHAPTER 5   SIMPLE MACHINES

Lever

A lever is one of the most elegant of the simple machines. All you need is a long beam of some sort that can pivot around a supporting point (the *fulcrum*). Levers are often put into three categories, or classes, depending on the arrangement of the fulcrum and load and of where force is applied. The model lever.scad has options to print any of the three classes. We first describe the classes and then the model.

Class 1 Levers

A *class 1 lever* has the fulcrum somewhere toward the center of the beam. Applying force downward on one side results in an upward force on the other. If the beam's length on either side of the fulcrum is equal, there is no mechanical advantage. However, it does change the force direction, thus allowing you to take advantage of gravity on one side to lift something on the other. A seesaw, for instance, is a class 1 lever. However, if one side is longer than the other, the mechanical advantage is the ratio of the lengths of the two sides. In principle, there is no limit to how long a lever can be, but it needs to be mechanically strong enough to lift the load without snapping or breaking the fulcrum support. Figure 5-2 shows a class 1 lever as it would have been oriented on the 3D printer, and Figure 5-3 with a load.

In Figure 5-3, the person's finger is showing where to apply force. Note that the class 1 lever has two indentations so that you can hang a weight at both ends if you want to. However, since the lever itself has finite (and unequal) mass on either side of the pivot and the pivot swings very freely, it is difficult to use this as a balance. It is best used to get a qualitative feel for the force to move an object attached to one end for a few different configurations. You might want to print out several different positions for the fulcrum and feel the difference in force for the same weight applied to the end of each.

CHAPTER 5   SIMPLE MACHINES

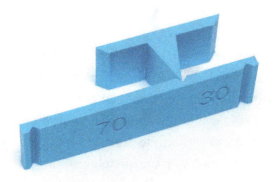

Figure 5-2. *Class 1 lever as oriented for printing*

Figure 5-3. *Class 1 lever*

Class 2 and 3 Levers

Other relationships apply to the other two classes of lever. Class 2 levers have a fulcrum on one end, and a load is toward the center of the beam. An upward force on the end opposite the fulcrum will lift the load.

CHAPTER 5   SIMPLE MACHINES

A *pry bar* is a metal bar that usually has a sharp bend at one end. When we use a pry bar, the first step is often using the sharp end as a wedge to get underneath or between things. Once the tip of the wedge is inserted, we might wiggle the bar up and down to help us push it in further. When we are trying to lift something to get the end of our pry bar further underneath, the bar is alternating between being used as a class 1 and class 2 lever.

When we push down on the end of the pry bar, the load is on the tip of the wedge, and the fulcrum is where the bend in the pry bar rests on the ground. Since the fulcrum is somewhere between the load and the force being applied, this is a class 1 lever. However, when we reverse the direction of force, the pry bar quickly begins to lift the load again. In this case, the tip of the pry bar has become the fulcrum, with the load now contacting a point further down the wedge.

Our ability to quickly reverse the input force while continuing to lift the load is what makes pry bars so effective. This is because we can take advantage of the brief period in between when little or no lever force is being applied to push the wedge further under the load.

A class 3 lever has the weight to be moved at one end and the fulcrum at the other. Force is applied at the marked spot partway along to pull up the load. Figure 5-4 shows a class 2 lever, and Figure 5-5 shows a class 3.

CHAPTER 5   SIMPLE MACHINES

When we think about the mechanical advantage of a class 1 lever, it is intuitive that the advantage is the ratio of the distance from the fulcrum to where force is applied, divided by the distance from the fulcrum to the load. That is also true for classes 2 and 3, but it is a little more subtle to think about.

In class 2 and 3 levers, the fulcrum is at one end, so either the force (class 2) or the load (class 3) part of the lever is the entire length. This means that class 2 levers always have a mechanical advantage greater than 1:1 and class 3 always worse than 1:1. The mechanical advantage of a class 2 lever is equal to the length of the entire lever, divided by the distance from the fulcrum to the load. The mechanical advantage of a class 3 lever is the inverse of this. Note that you must hold the class 3 lever down at the fulcrum, too, in this design.

Class 3 levers may seem useless since their mechanical advantage is always less than 1. However, applying more force is not always what you want. With a mechanical advantage less than one, you multiply the distance (and thus the speed) that the load moves. This is how muscles like the bicep actuate joints in the human body. Baseball bats and similar stick-shaped sporting equipment all take advantage of this speed amplification created by having a mechanical advantage less than 1.

The model in Listing 5-2 creates a lever and fulcrum. Table 5-2 describes the variables you can change to try out different configurations.

CHAPTER 5 SIMPLE MACHINES

Figure 5-4. Class 2 lever

Caution The lever STL file is laid out as two pieces—the lever and the fulcrum—oriented as they would be when used with each other. You may want to separate and reorient the parts in your slicing software to make more efficient use of the space on your printer's platform.

165

CHAPTER 5   SIMPLE MACHINES

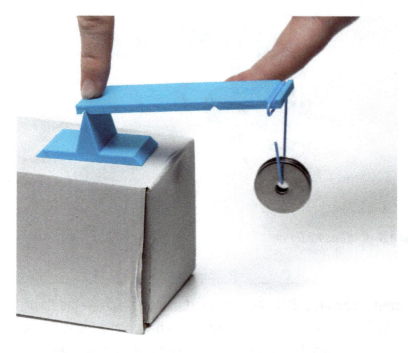

Figure 5-5. *Class 3 lever*

Table 5-2. *Lever Variables*

Variable	Default Value and Units	Meaning
lever[a,b]	lever[70, 30] mm	Length of the lever on either side of fulcrum (class 1) or load/resistance application point (class 2 or 3)
class	1, 2, or 3	Type of lever
width	30 mm	Width of lever and fulcrum

Listing 5-2. Lever Model (file lever.scad)

```
//OpenSCAD model to creates all three classes of lever
//File lever.scad
//(c) 2016-2024 Rich Cameron
//for the book 3D Printed Science projects, Volume 1
//Licensed under a Creative Commons, Attribution,
//CC-BY 4.0 international license, per
//https://creativecommons.org/licenses/by/4.0/
//Attribute to Rich Cameron, at
//repository github.com/whosawhatsis/3DP-Science-Projects

//two lengths of the lever
lever = [70, 30];
//class of lever
class = 1; //[1, 2, 3]
//height of point of fulcrum, in mm
fulcrum_height = 30;
//width in mm
width = 30;

echo(str("mechanical advantage: ",
  (class == 1) ? str(lever[0], ":", lever[1]) :
  (class == 2) ? str(lever[0] + lever[1], ":", lever[1]) :
  (class == 3) ? str(lever[0], ":", lever[0] + lever[1]) :
  0
));

linear_extrude(width, convexity = 5) difference() {
  intersection() {
    square(fulcrum_height * 2, center = true);
    rotate(-135) square(fulcrum_height * sqrt(2));
  }
```

CHAPTER 5   SIMPLE MACHINES

```
    difference() {
      square(fulcrum_height * 2 - 10, center = true);
      intersection_for(a = [1, -1]) rotate(-135 + a * (45 - 15))
        square(fulcrum_height * sqrt(2));
    }
}

difference() {
  linear_extrude(width, convexity = 5) difference() {
    translate([-5 - ((class == 1) ? lever[1] : 0), 0, 0])
      square([10 + lever[0] + lever[1], 5]);
    translate([0, 2.5, 0]) rotate(-135) square(5);
    translate([lever[0], 2.5, 0])
      rotate(((class == 3) ? -135 : 45)) square(5);
    translate([
      ((class == 1) ? -lever[1] : lever[0] + lever[1]),
      2.5,
      0
    ]) rotate(((class == 2) ? -135 : 45)) square(5);
  }
  translate([lever[0] / 2, 5, width / 2])
    rotate([90, 0, 180])
      linear_extrude(height = 1, center = true, convexity = 5)
        text(
          str(lever[0]),
          size = min(width / 3,
          lever[0] / 3),
          halign = "center",
          valign = "center"
        );
  translate([
```

```
        (class == 1) ? -lever[1] / 2 : lever[0] + lever[1] / 2,
        5,
        width / 2
    ]) rotate([90, 0, 180])
        linear_extrude(height = 1, center = true, convexity = 5)
            text(
                str(lever[1]),
                size = min(width / 3,
                lever[1] / 3),
                halign = "center",
                valign = "center"
            );
}
```

Screw

We usually think of a *screw* as something that holds things together. More generally, though, turning a screw takes rotational motion and changes it into linear motion along the screw's axis. 3D printers often use a screw to drive one or more of their axes.

Screws used as fasteners (e.g., machine screws with nuts, shown in Figure 5-6) are designed to use mechanical advantage to convert rotational force into tensile and compressive forces to produce friction to hold things in place. They have a helical groove wrapped around them called the *thread*.

Wood screws (a compound machine, consisting of a screw and a wedge) intentionally increase this friction by wedging apart the wood as they are inserted. The compression of the wood around the screw produces a radial force against the screw that multiplies the friction. This helps hold the screw in place. Machine screws, on the other hand, are screwed into a nut or into another preexisting hole with matching threads. Machine screws fasten things together by compressing them between these threads and the head of the screw.

CHAPTER 5 SIMPLE MACHINES

The mechanical advantage of a screw is proportional to how far apart the threads are (the *pitch*); the closer they are, the higher the advantage. However, screws are usually far off the ideal for most practical purposes because there is so much friction on a screw embedded in a material. Wikipedia has a very good brief explanation in the article "Screw mechanism."

Figure 5-6. *Machine screws with nuts*

The model archimedesScrew.scad creates an *Archimedes' screw*, sometimes also called an *Egyptian screw*. This machine was used in antiquity to raise water up from one level to another. The Greek philosopher Archimedes is reported to have seen one in Egypt about 2,400 years ago, and since he wrote about the device, it is often attributed to him. Figure 5-7 shows the print.

To raise water (or other substances) with the screw, the screw is tilted a bit to keep the water in the screw's threads. Then as it is turned, water will rise up the inclined plane (Figure 5-8). Create one and try it out (but see the printing tip that follows). Table 5-3 gives the variables in the model, which follows in Listing 5-3.

CHAPTER 5 SIMPLE MACHINES

Tip The point where the screw joins the base is fragile. Take it off the 3D printer platform by flexing the platform, if your printer allows that, or by holding it by the base, not by the screw thread.

Figure 5-7. Archimedes screw print

CHAPTER 5   SIMPLE MACHINES

Figure 5-8. Raising water with the screw

Table 5-3. Archimedes Screw Model Variables

Variable	Default Value and Units	Meaning
pitch	16 mm	How far up the screw steps with each turn
r	12 mm	Radius of the screw. Total height will be a little more than this times pitch
wall	0.8 mm	Wall thickness, ideally double your nozzle diameter
turns	5	Number of turns
fn	180	Number of segments per turn

Listing 5-3. Screw Model (file archimedesScrew.scad)

```
//OpenSCAD model to create an archimedes screw
//File archimedesScrew.scad
//(c) 2024 Rich Cameron
//for the book 3D Printed Science projects, Volume 1
//Licensed under a Creative Commons, Attribution,
//CC-BY 4.0 international license, per
//https://creativecommons.org/licenses/by/4.0/
//Attribute to Rich Cameron, at
//repository github.com/whosawhatsis/3DP-Science-Projects

//how far up the screw steps with each turn (mm)
pitch = 16;
//radius of the screw (mm)
r = 12;
//wall thickness, ideally double your nozzle diameter (mm)
wall = .8;
//number of turns
turns = 5;
//number of segments per turn
fn = 180;

$fs = .2;
$fa = 2;

rawthread = [
  r * [0.01, sin(-45) * 2],
  for(a = [-45:2:45]) r * [cos(a), sin(a)],
  for(a = [45:-2:-45]) (r - wall) * [cos(a), sin(a)],
  r * [0.01, sin(-45)],
];

%polygon(rawthread);
```

CHAPTER 5   SIMPLE MACHINES

```
points = [for(a = [0:360 / fn:turns * 360])
  for(slice = rawthread) [
    slice[0] * sin(a),
    slice[0] * cos(a),
    a * pitch / 360 + slice[1]
  ]
];
*for(i = points) translate(i) cube(.2, center = true);
translate([0, 0, r * sin(45) * 2]) polyhedron(points, [
  for(i = [0:len(rawthread) / 2 - 2])
    [i, i + 1, len(rawthread) - 1 - i],
  for(i = [0:len(rawthread) / 2 - 2])
    [i + 1, len(rawthread) - 2 - i, len(rawthread) - 1 - i],
  for(i = [0:len(rawthread) / 2 - 2])
    [i + 1, i, len(rawthread) - 1 - i]
      + [
        len(points) - len(rawthread),
        len(points) - len(rawthread),
        len(points) - len(rawthread)
      ],
  for(i = [0:len(rawthread) / 2 - 2])
    [i + 1, len(rawthread) - 1 - i, len(rawthread) - 2 - i]
      + [
        len(points) - len(rawthread),
        len(points) - len(rawthread),
        len(points) - len(rawthread)
      ],
  for(i = [0:len(points) - len(rawthread) - 1])
    [i + 1, i, i + len(rawthread)],
  for(i = [0:len(points) - len(rawthread) - 1])
```

```
    [i, i + len(rawthread) - 1, i + len(rawthread)],
]);

cylinder(r = r, h = r / sqrt(2) + wall);
cylinder(r = r * .02, h = r / sqrt(2) + pitch * turns);
```

Wheel and Axle

The first known instances of a *wheel and axle* were in Mesopotamia between 5,000 and 6,000 years ago. A wheel makes it easier to move things by allowing a small motion at the hub to be translated to a larger motion at the rim, creating mechanical advantage. Wheels also reduce frictional resistance with the ground by having a small contact area with the ground; friction in a wheel occurs at the axle. Wheels also function like a variable-angle inclined plane at their leading and trailing edges as they roll over the ground, smoothing out any bumps in the surface they roll over.

Some ancient civilizations (including the Egyptians) managed engineering feats without the wheel or the pulley. In modernity, wheels are usually seen in combination with other simple machines, like the combination of lever and wheel and axle we see in the humble wheelbarrow (Figure 5-9). The wheel allows the user to move around heavy loads, and the handles (levers) allow hoisting the barrow up onto its wheel. (Otherwise, it rests on its feet.)

We have created a 3D printable wheel that fits on a bamboo skewer (the kind used for kebabs) or another axle of your choice, which works well to make a simple toy chassis for experiments. The model `skewerWheel.scad` by default creates a set of 4 wheels. To create a toy car platform as a base for wheel and axle experiments, take four craft sticks and tape them together. Then cut a straw into two lengths a little longer than the width of the taped-together craft sticks (Figure 5-10), and tape those lengths perpendicular to the straw (Figure 5-11).

CHAPTER 5   SIMPLE MACHINES

Figure 5-9. Wheel and axle, combined with a lever, in a modern wheelbarrow

Figure 5-10. Making the wheel and axle platform

CHAPTER 5 SIMPLE MACHINES

Figure 5-11. *Assembled platform from below*

Next, cut two identical lengths of bamboo skewer about 20 mm longer than the straws, cutting off the pointy ends. Put a wheel on one end of each one. Note that the patterned side of each wheel faces inward toward the skewer and center of the chassis since it is designed to flex a bit so the skewer will fit. Now pass the skewer through the one straw and attach a wheel to the other side. Do the same for the other straw. You now have a nice wheel and axle platform for experiments (Figure 5-12). The wheels should turn freely and can be used as a base to slide down an inclined surface and to see how much wheels reduce sliding friction. Table 5-4 and Listing 5-4 give the model details.

CHAPTER 5   SIMPLE MACHINES

Figure 5-12. *Assembled platform on its wheels*

Table 5-4. *Wheel Model Variables*

Variable	Default Value and Units	Meaning
id	3.1 mm	Size of skewers (mm), should be a friction fit
d	25 mm	Wheel diameter
wall	0.8 mm	Minimum wall thickness (should be double your nozzle diameter)

Listing 5-4. Wheel Model (file skewerWheel.scad)

```
//OpenSCAD model to create wheels to fit on a bamboo skewer
//File skewerWheels.scad
//(c) 2024 Rich Cameron
//for the book 3D Printed Science projects, Volume 1
//Licensed under a Creative Commons, Attribution,
```

CHAPTER 5 SIMPLE MACHINES

```
//CC-BY 4.0 international license, per
//https://creativecommons.org/licenses/by/4.0/
//Attribute to Rich Cameron, at
//repository github.com/whosawhatsis/3DP-Science-Projects

//size of skewers (mm). should be a friction fit
id = 3.1;
//wheel diameter
d = 25;
//minimum wall thickness (should be double your nozzle size)
wall = .8;

hub = d - wall * 6;
$fs = .2;
$fa = 1;

for(x = [1, -1], y = [1, -1])
  translate((d / 2 + 1) * [x, y, 0]) {
    linear_extrude(6) offset(wall / 3) offset(-wall / 3)
      difference() {
        circle(d / 2);
        circle(id / 2);
        for(a = [0:120:359]) rotate(a) {
          translate([0, -1, 0])
            square([hub / 2 - wall, id / 4]);
          difference() {
            circle(hub / 2);
            circle(id / 2 + wall);
            for(a = [0:120:359]) rotate(a)
              translate([0, -1, 0]) mirror([0, 1, 0])
                square([hub / 2 + wall, wall]);
          }
        }
      }
```

CHAPTER 5   SIMPLE MACHINES

```
    }
    linear_extrude(.4) difference() {
      circle(d / 2);
      circle(id / 2 + wall);
    }
  }
}
```

We have also included a 3D printable chassis (model `chassis.scad`) that can be used instead of popsicle sticks and straws. The variables for the model are in Table 5-5 and the model in Listing 5-5. This can be assembled with skewers and printed wheels just like the popsicle stick version. Alternatively, you can obtain radial roller bearings with a bore matching the printed part, along with matching screws, to build a version that rolls even better. Figure 5-13 shows it assembled with wheels and skewers.

Figure 5-13. *Assembling chassis and wheels*

CHAPTER 5   SIMPLE MACHINES

Table 5-5. Variables in chassis.scad

Variable	Default Value and Units	Meaning
size	[40, 100] mm	[x, y] dimensions
wheelbase	80 mm	Distance between axles, should be less than y dimension
hole	4 mm	Hole diameter, should fit skewer axles very loosely

When using the printed wheels with a skewer, the skewer forms a *live axle*, which is an axle that spins with the wheel. Since each pair of wheels shares a live axle, the wheels will spin together. The friction in this system is between the axle and the chassis. In contrast, the version using bearings employs screws as *dead axles*. These axles do not spin along with the wheels, and each wheel will rotate independently of the others. However, when rolling along the ground, their rotation will mostly synchronize due to friction with the ground.

The radial bearings that we are using as wheels have rolling balls inside that reduce the friction between their inward- and outward-facing surfaces to nearly zero. In most real-world applications, a live axle would be mounted to a chassis using similar bearings.

Listing 5-5. Chassis Model (file `chassis.scad`)

```
//OpenSCAD model of a printable chassis for skewerWheels.scad
//File chassis.scad
//(c) 2024 Rich Cameron
//for the book 3D Printed Science projects, Volume 1
//Licensed under a Creative Commons, Attribution,
//CC-BY 4.0 international license, per
```

CHAPTER 5   SIMPLE MACHINES

```
//https://creativecommons.org/licenses/by/4.0/
//Attribute to Rich Cameron, at
//repository github.com/whosawhatsis/3DP-Science-Projects

size = [40, 100];
wheelbase = 80;
hole = 4;

$fs = .2;
$fa = 2;

difference() {
  union() {
    linear_extrude(hole + 2, center = true)
      offset(2) offset(-2) square(size, center = true);
    for(i = [-1, 1]) translate(i * [0, wheelbase / 2, 0])
      rotate([0, 90, 0])
        linear_extrude(size[0] + 2, center = true)
          intersection() {
            hull() {
              circle(hole / 2 + 1);
              scale([sqrt(2), .5])
                circle(hole / 2 + 1, $fn = 4);
            }
            square(hole + 2, center = true);
          }
  }
  for(i = [-1, 1]) translate(i * [0, wheelbase / 2, 0])
    rotate([0, 90, 0])
      linear_extrude(size[0] + 3, center = true)
        circle(hole / 2);
}
```

Pulley

A *pulley* is a wheel and axle with a groove around the wheel's rim. A rope, chain, or belt of some sort goes into the groove and is used to move something attached to it. In the simplest form, a pulley just changes the direction of a force, but not its magnitude. The idler pulley in Figure 5-14, for example, changes the direction of the cable running over it.

Figure 5-14. Idler pulley

If more than one pulley is used, then the weight of an object is distributed over several sections of rope. A block and tackle and related, more complex devices exploit this feature. The mechanical advantage depends on the number of pulleys, their distribution, and how the rope is attached—look up "block and tackle" to see some examples (such as the Wikipedia article, "Block and tackle"). The diameter of a pulley typically does not affect its mechanical advantage. The exception is when a pulley's axle is attached to something else, such as another pulley with a different diameter, so that the two turn together.

CHAPTER 5   SIMPLE MACHINES

The model `pulley.scad` creates a set of grooved wheels and a frame (Figure 5-14). The variables you can adjust are in Table 5-6, and the model is in Listing 5-6. These can then be assembled (once again using a skewer for an axle) either into a single pulley (Figure 5-15), or you can print a pair to assemble a block and tackle (Figure 5-16).

To assemble the block and tackle, print and assemble two sets of parts from `pulley.scad` as shown in Figure 5-16. You can either print the wheels of the pulley as shown in the figure, or you can purchase a suitable-sized bearing to snap into it (to be between the 3D printed wheel and the skewer axle). If you add a purchased bearing, dimensions so the 3D printed part will fit around it will need to be adjusted, as shown in Table 5-6. You can also use grooved bearings (size U624ZZ, for example) so that you do not need the printed sheave around the bearing.

You can use multiple pulleys to assemble a block and tackle. Figure 5-16 shows a simple example, but a block and tackle often uses pulleys with two or more sheaves each, so that rope can be looped through them multiple times, for even greater mechanical advantage. A block and tackle creates mechanical advantage by running a single rope back and forth between a set of pulleys multiple times.

With a single pulley, as shown in Figure 5-15, pulling one unit of the length of the rope on one side of the pulley reduces the length on the other side by an equal amount. When using a block and tackle, you reduce the total length of several parallel stretches of rope by that amount. Since the load is distributed across these parallel stretches of rope, the distance that the load travels is divided by the number of them. The simple block and tackle in Figure 5-16 has two such stretches of rope, so the mechanical advantage is 2:1.

CHAPTER 5   SIMPLE MACHINES

Figure 5-15. *Single pulley*

Figure 5-16. *Block and tackle*

Table 5-6. Pulley Variables

Variable	Default Value	Meaning
bearing_d	4 mm	Outer diameter of bearings. If not using bearings, just make this about 1 mm more than bore to ensure a loose fit.
sheave_d	32 mm	Outer diameter of the wheel or "sheave" going around the bearing. Set to zero to print none (if using grooved bearings, for example).
bore	3.2 mm	This should fit your skewer or other axle tightly enough to keep it from slipping out.
width	7 mm	Width of bearings, should be wider than your string.
clearance	1 mm	Extra space to allow things to spin freely. Can be small if using bearings (depending on printer accuracy).
bearing_fit	0.1 mm	Clearance between bearing and sheave. Should be a fairly tight fit.
n	1	Number of pulleys per block. Higher numbers can be used to make a block and tackle with more mechanical advantage.
inner_wall	1 mm	Width of inner wall between pulleys. This should be about twice your nozzle diameter. Ignored if n = 1. Strings should be thicker than inner_wall + clearance to ensure that they do not slip between pulleys.
outer_wall	2 mm	Width of the walls around the outside of the pulley block.

CHAPTER 5 SIMPLE MACHINES

Listing 5-6. Pulley Model (file pulley.scad)

```
//OpenSCAD model for making pulleys/blocks and tackles
//File pulley.scad
//(c) 2024 Rich Cameron
//for the book 3D Printed Science projects, Volume 1
//Licensed under a Creative Commons, Attribution,
//CC-BY 4.0 international license, per
//https://creativecommons.org/licenses/by/4.0/
//Attribute to Rich Cameron, at
//repository github.com/whosawhatsis/3DP-Science-Projects

bearing_d = 4;
sheave_d = 32;
bore = 3.2;
width = 7;
clearance = 1;
bearing_fit = .1;
n = 1;
inner_wall = 1;
outer_wall = 2;

$fs = .2;
$fa = 2;

translate([0, 0, bore / 2 + outer_wall]) rotate([0, -90, 0])
  difference() {
    union() {
      linear_extrude(
        (width + clearance) * n
        + inner_wall * (n - 1)
        + outer_wall * 2,
        convexity = 10
```

187

CHAPTER 5 SIMPLE MACHINES

```
    ) difference() {
      intersection() {
        circle(max(
          bearing_d,
          sheave_d) / 2 + width + outer_wall
        );
        square([bore + outer_wall * 2, 1000], center = true);
      }
      circle(bore / 2);
    }
    for(i = [-1, 1]) translate([
      0,
      i * (max(
        bearing_d,
        sheave_d
      ) / 2 + width + 3 + outer_wall),
      (
        (width + clearance) * n
        + inner_wall * (n - 1)
        + outer_wall * 2
      ) / 2
    ]) difference() {
      rotate([0, 90, 0]) linear_extrude(
        bore + outer_wall * 2,
        center = true,
        convexity = 5
      ) difference() {
        circle(3 + outer_wall);
        circle(3);
      }
      rotate(i * [-90, 0, 0])
```

```
      linear_extrude(100) for(j = [1, -1])
        hull() for(k = [0, 1])
          translate(j * [
            3 + outer_wall / 2 + 10 * k,
            0,
            0
          ])
            circle(3);
  }
}
for(i = [0:n - 1])
  translate([
    0,
    0,
    i * (width + clearance + inner_wall) + outer_wall
  ]) {
    rotate_extrude() union() {
      translate([bore / 2 + inner_wall, 0, 0])
        square([
          max(
            bearing_d,
            sheave_d
          ) / 2 + width - bore / 2 - inner_wall,
          width + clearance
        ]);
      translate([0, clearance / 4, 0])
        square([
          bore / 2 + inner_wall + 1,
          width + clearance / 2
        ]);
    }
```

CHAPTER 5   SIMPLE MACHINES

```
      }
   }
for(i = [0:n - 1])
  translate([
    sheave_d / 2 + 1,
    i * (max(bearing_d, sheave_d) + 1),
    0]
  ) rotate_extrude()
    translate([bearing_d / 2 + bearing_fit, 0, 0])
      difference() {
        square([
          (sheave_d - bearing_d) / 2 - bearing_fit,
          width
        ]);
        translate([
          width / sqrt(2) + max(
            1,
            (sheave_d - bearing_d
          ) / 2
          - bearing_fit
          - width / sqrt(2) * (1 - sin(45))),
          width / 2,
          0
        ])
          circle(width / sqrt(2));
      }
```

Printing Suggestions

These models can be scaled up (made bigger), but only the wedge and inclined plane can safely be made smaller. The other models may have tolerance issues if they are made too small, and any non-printed parts that they are designed to interface with obviously will not fit if the model is scaled. Generally, it will be better to change the scaling variables in OpenSCAD than to use 3D printer software scaling functions.

> **THINKING ABOUT THESE MODELS: LEARNING LIKE A MAKER**
>
> Unlike many of the other models in this book, these did not start with an abstraction. Rather, they are prints of things that are normally 3D objects. While creating them, we did not struggle to think about how to represent them since the objects were physical already. Here, the challenge is that these building blocks are normally part of a more complex whole. You will need to spend some time thinking about what you want to learn (or teach) from these models to decide which ones to print out and how to arrange them to do something useful or to demonstrate a principle.
>
> Mechanical advantage can, for the most part, be shown with a string taped to a weight (a few coins, possibly). We encourage you to try these out and see what you can construct.

Where to Learn More

Since simple machines are a required topic in many elementary and middle school curricula, there are correspondingly large numbers of online games and student and teacher support materials available online. Try searching for "simple machine games."

CHAPTER 5   SIMPLE MACHINES

The definition of *simple machines* is a little arbitrary. Different disciplines have varying opinions about what the simple machines include. Sometimes gears are included as a distinct seventh simple machine; others feel that meshing gears are really a wheel and axle combined with wedges. Either way, there are already a lot of fun gears available to 3D print. For example, search "gears" on `www.printables.com`, but be careful to check the comments to see if the design has been proven out. Many of these designs will be challenging prints that require a well-tuned printer. In Chapter 9, we look at more complex but still very fundamental mechanisms that build on these machines.

Teacher Tips

Since simple machines are so ubiquitous, one simple thing to do is to have students look around at common household or classroom objects and deconstruct them to find the simple machines. Kitchen hand tools are particularly rich sources of machines. These everyday hand tools can be surprisingly complex. Then you can have students design their own combinations of simple machines to solve some theme or problem.

Some schools have given students projects to recreate historic machines, sometimes with a twist. Generally, the topics around simple machines are covered in the NGSS Forces and Interactions standards (MS-PS-2, `www.nextgenscience.org/msps2-motion-stability-forces-interactions`) and Energy standards (MS-PS-3, `http://www.nextgenscience.org/msps3-energy`).

Science Fair Project Ideas

When these simple machines meet up with the real world, nonideal effects like friction come into play. You might characterize some of these real-world effects and find a systematic way to manage them to minimize inefficiency. Analyze what existing products do to be efficient, and think about what you might change to do something better.

Getting a little more into product design issues, you might think about how machines meant for humans to use without additional power (hand tools, daily-use objects) might be more or less efficient for people with different size hands or in different environments (e.g., wearing gloves or not). You could also investigate what kinds of things make these simple machines break or not work very well and what enhancements are the best shields against that.

Some good books on both good user experience design and the origins of classic mechanisms are Donald Norman's *The Design of Everyday Things* (Basic Paperbacks, 2002) and Henry Petroski's books on how things fail when good design principles are violated. Petroski's classic work is *To Engineer Is Human: The Role of Failure in Successful Design* (Vintage, 1992), but he also has entire books focused on a few inventions, such as the toothpick.

Summary

In this chapter, we learned about how to 3D print six simple machines: wheel and axle, pulley, inclined plane, wedge, lever, and screw. We covered a bit of the history of these machines, how they generate mechanical advantage, and how to print them. We wound up with some ideas on how to demonstrate the idea of a simple machine with historical recreations or perhaps some research into good product design.

CHAPTER 6

Plants and Their Ecosystems

When we think about plants, dynamic is not usually a word that comes to mind. Plants are part of the landscape. Unless you are a gardener, you might never give the trees and smaller plants around you a second thought. The fact that plants are, well, planted in place has some evolutionary implications, though. If you are a life form that cannot move (other than by growing vertically or spreading horizontally), and the environment changes, your offspring have the options of adapting or dying out. This means that plants have developed a wild array of forms to follow their ecological niche function.

    This chapter develops a model of how plants grow to take in as much sunlight as possible and to manage water and other resources and how this function might need to vary in different ecosystems. Our intent is that you will play with the model and create yourself some plant communities to print out and think about. This chapter can therefore be approached at many different levels: from creating flower and plant models for elementary students to thinking about "evolving" a plant for an ecosystem or to looking at the mathematics underlying the model.

CHAPTER 6   PLANTS AND THEIR ECOSYSTEMS

MODELS USED IN THIS CHAPTER

This chapter uses two different OpenSCAD models. For more on 3D printing and creating and using math models in OpenSCAD, check out Chapter 1. Select 3D printable STL example files are included in the repository as well, and we will describe them in the chapter text. The OpenSCAD models are

- FlowersPlants.scad: This model produces flowers and plants, as described in the chapter. It is included as Listing 6-1.
- DripLeaf.scad: This model produces leaves for a jungle plant (base made with the Listing 6-1 model). It is included as Listing 6-2.

To do one experiment noted in this chapter, you will need to have access to bright sunlight or a good flashlight.

Botany Background

Botany, the study of plants, is one of the oldest sciences. Everyone who either forages for or farms plants will do better the more they have predictive power about what plant will grow where and how to encourage plants that have the best potential as a food source.

If an animal's habitat changes, it can adapt (on evolutionary time scales of millennia) or migrate somewhere else. Plants, however, cannot move in the same way. They can hitchhike to another climate by spreading seeds, but often a species that does not adapt fast enough will just die out.

Wikipedia says there are somewhere between 300,000 and 315,000 species of plants (see Wikipedia's "Plant" article). However, because of ecosystem destruction coupled with climate changing rapidly, the rate of

CHAPTER 6   PLANTS AND THEIR ECOSYSTEMS

extinction is very high. Just how high is a matter of some intense debate, but most reports assume double-digit percentages of plants may be gone in the next century.

Add to that many invasive species, which are spread to new habitats accidentally or deliberately as people jet around the planet. These factors mean that it is important to think about plants as part of an overall ecosystem and consider how the plants and resources in an area play together.

To thrive, plants need six things: light, water, gasses, heat, nutrients, and mechanical support. We are going to focus on just two of those—light and water—and talk about how plants fit into environments where one of those is limited.

Water

Plants need water both for their chemical reactions and most of their structures. However, plants that receive too much water can rot. Therefore, plants in wet environments like jungles and rainforests have developed strategies to get rid of water, like waxy leaves that come to a point to drip away water.

On the other hand, desert plants work hard to both hang on to water and to keep animals and insects from chomping on the structures they managed to create with their hard-won water. Desert plants often have a milky, acid sap (notably plants in the genus *Euphorbia*, commonly called spurge). This sap will squirt out on anything that takes a bite out of a euphorbia to get some moisture, presumably discouraging it from trying it again. Other desert plants have spines or spikes to discourage nibbling, like the agave in Figure 6-1, as well as tough outer surfaces that hold water in under a desert summer sun.

CHAPTER 6   PLANTS AND THEIR ECOSYSTEMS

Figure 6-1. *Agave spines and tough outer surface defend their hard-earned moisture*

Sunlight

Plants, for the most part, make their living by capturing the sun's energy in chemical bonds. Too little sun, and a plant may fail to thrive and reproduce. However, some plants cannot handle too much sun and have evolved to live in the shadow of other plants. Plants can get sunburned, too.

CHAPTER 6 PLANTS AND THEIR ECOSYSTEMS

Nutrients

Nitrogen is a very abundant element. Plants need nitrogen to grow and typically get it from the soil. Plants are also dependent to one degree or another on many other elements, such as potassium and phosphorus, for different functions. Some plants put needed elements back in the soil so that other plants can use them (like "nitrogen-fixing" plants such as fava beans). Others die if the soil has too high a concentration of elements that might be needed for another plant.

If you are a gardener and buy fertilizer, you will often see it labeled with numbers like 24-8-16. The first number (in this case, 24) is the percentage of nitrogen by weight in the preparation. The second (8, here) is the percentage of available phosphate, and the third (16 in our example) represents soluble potash (potassium in water-soluble form, often written K_2O, a sort of abbreviation for its chemical representation). These are usually called NPK values since K is the chemical symbol for the element potassium. Different fertilizers are sold because different plants may need different proportions of nutrients from each other. Even the same plant may need different fertilizers at different points in its life cycle and may need trace nutrients too.

Plant Communities

When you create a garden, you need to consider whether the chemistry, water, and sun needs of the plants you put next to each other are compatible. In natural environments, evolution takes care of which plants survive and which do not, which brings us to the topic of how plants form communities. Figure 6-2 shows a typical cooperative ecosystem of *Camellia japonica* bushes and trees in the *Cocculus* genus.

Camellia japonica evolved in a cool climate in Japan and were transplanted to the Southwestern United States, where they flower in winter in the shade of other trees. Its relative, *Camellia sinensis*, which means "camellias from China," is known best as the plant that gives us tea. The trees above get all the sun they want, and the camellias flourish in their shade and leaf litter. (If camellias are out in the sun in Southern California in the summer, the leaves get blotchy sunburn where they are exposed.) The *Cocculus* tree is native to the Himalayas. Together, the camellias and cocculus thrive in Southern California's Mediterranean climate.

Figure 6-2. *Camellia japonica bushes shaded by Cocculus trees in February, Pasadena, California*

CHAPTER 6   PLANTS AND THEIR ECOSYSTEMS

This is an example of how plants compete for resources like sun but also may cooperate. A plant (like these camellias) that cannot handle much sun or wind might have the best chance of survival under a tree. Some plants might need different nutrients than others and so can grow close together without depleting the soil—one plant might even deposit nutrients another needs. Some plants deposit chemicals in the soil that prevent other plants from germinating. Others try to outdo each other with attractive flowers to attract more pollinating insects and birds.

All plants grow and reproduce, either in competition or cooperation with one another, in the soil in which they sprouted. As the environment around them changes, they adapt or die out. In this chapter, we show some of the mathematics of plant growth and develop some simple models of plant leaves and flowers. Then you can think about how you might use these models to think about an entire ecosystem and how it is affected when things change.

The Mathematics of Plant Growth

When a plant adds on new growth, usually the new material is layered on at a part of the plant called the *meristem*. It would be a waste of the plant's energy and biomass if the leaves were mostly covered by other leaves or if all the petals of a flower (meant to attract pollinators) were in a big pile rather than spread out as broadly as possible to let bees and hummingbirds know that the buffet is on the table.

Evolution has favored plants that have the most efficient distribution of leaves and flower petals. In many cases, this efficient distribution spaces subsequent leaves or petals by the *golden angle* (about 137.52 degrees). The story of this angle is closely related to numbers called the *Fibonacci sequence*. We talk about all that in the next sections and then use this model in the rest of the chapter as the basis for our 3D printed plant models.

CHAPTER 6   PLANTS AND THEIR ECOSYSTEMS

The Golden Ratio

Two integer numbers, *a*, and *b*, with *b* the larger number, are said to be in the golden ratio to each other if

$$\frac{a+b}{b} = \frac{b}{a}$$

If that is true, then the ratio of *a*/*b* is equal to the golden ratio. The golden ratio is exactly equal to

$$\phi = \frac{1}{2}\left(1 + \sqrt{5}\right)$$

which is approximately 1.618. It is usually referred to by the Greek letter ϕ (phi), most commonly pronounced "fie" in the English-speaking world, but "fee" in Greek and by some mathematicians and physicists. It has some other interesting properties, like this recursive definition:

$$\phi = 1 + \frac{1}{\phi}$$

If you have seen the quadratic formula, you will see that you can multiply this equation through by ϕ, wind up with a quadratic, and find its root to get the solution in terms of square root of 5 given earlier.

Phi is an irrational number (like pi), which means that you can never write down all its digits. There is no pattern to its digits either, as there is for, say, 2/3. The ancient Greeks and Egyptians knew about it and used it in art (to scale the sides of the pyramids, for example), since the proportion 1:1.618 feels pleasing to the eye. There is a good extensive explanation with pictures at www.mathsisfun.com/numbers/golden-ratio.html.

The Golden Angle

Suppose you went around a circle 1.618 (or ϕ) times. There are 360 degrees in a circle, so that would be 584.2 degrees. If you go 360 degrees, or 720 degrees, you wind up back where you started. Suppose we subtract 584.2 degrees from 720 degrees to see how short we are from going around twice. That winds up being roughly 137.52 degrees, which is called the golden angle.

If you want to lay out something around a circle such that every time you go around you put something down a little different place than the last time, one way to ensure that is to space things out by multiples of the golden angle. Because it is not possible to write down the golden angle exactly (since it is an irrational number), each subsequent placement will be just a bit off in relation to its predecessors. Plants have evolved in many cases to lay out their material in ways that are very similar to this, as we will see shortly.

Fibonacci Sequence

Closely related to the concept of the golden ratio is the Fibonacci sequence, credited to the Italian mathematician Fibonacci, who published them in 1212 to try and model the breeding of rabbits. He failed in that, since populations follow a different relationship, but nevertheless laid the groundwork for a lot of other mathematics.

The Fibonacci sequence is defined such that it begins either with {0, 1} or {1, 1} (depending on whom you ask) and that if you add any two consecutive numbers in the sequence, you get the next number. The sequence is {(0,) 1, 1, 2, 3, 5, 8, 13, 21, 34, 55 . . .}. If you divide a number in the sequence by the prior one, you get {1/1, 2/1, 3/2, 8/5, 13/8, 21/13, 34/21, 55/34 . . .} or {1, 2, 1.5, 1.600, 1.625, 1.615, 1.619, 1.618 . . .}, which as you can see is getting closer and closer to the golden ratio, which is a bit larger than 1.618.

Mathematicians have shown that the ratio of consecutive numbers in this sequence approaches the golden ratio as the numbers get bigger and bigger. You can read the Wikipedia "Fibonacci number" article if you want to learn more.

Phyllotaxis

The process of *phyllotaxis* generates leaves in a growing plant. Leaves are where the plant makes a living—taking sunlight and water and creating the sugars the plant needs to live and reproduce. Leaves are generated at the plant's meristem, usually at the tip of a stem or branch, which has specialized cells that produce new plant material.

If the leaves were regularly spaced (say at 90-degree angles), then at least some of the leaves would overlap. What you really want (if you are a plant) is to have a sequence that never quite repeats and places leaves evenly around your center. As we have seen earlier in the chapter, many plants have a number of leaves or flower petals corresponding to the Fibonacci sequence of numbers (many with 5 or 8 leaves or petals) with subsequent leaves spread out at the golden angle.

If the leaves are pushing out of a center and forming new leaves that push out old ones (or form on the outside of the old ones), a spiral pattern arises. There are various spiral patterns in nature, including Archimedean spirals and logarithmic ones. In our model here, we are implementing a very simplistic spiral with a linearly increasing radius to give an easy-to-describe starting point for your explorations.

The Models

The models in this chapter attempt to generate somewhat stylized plants and flowers following the mathematics in the preceding section. We have developed two different OpenSCAD models. One can generate desert plants and flowers, and the other is an add-on to the first to enable plants like jungle plants with big leaves and long stems. The models have many parameters to govern how the leaves or petals are angled and spaced but maintain the overall spiral arrangement (and use a Fibonacci number of petals or leaves). As we talk about in the "Thinking About the Models: Learning Like a Maker" section later, a relatively simple rule can generate complex behavior.

To generate each leaf/petal, we recycled some of the concepts used to create the airfoil cross sections in Chapter 4. You will see some strong similarities in the code in this chapter and the basic airfoil simulation there. Next, we tried to start with botanical first principles and think about what kind of plant would likely do well in a particular environment—and then see how close our plants were to real ones in the ecosystem.

We think that these models may be a fun way to focus discussion on how plant forms evolve based on evolutionary pressures. Would a mutation that produced a cactus do well in a cold, rainy climate? Probably not. You might think about designing a plant community and then trying to find and model the most disruptive invasive species that could possibly upend this community.

For floral simulations, you might think about what type of pollinator is present in the environment and create an arrangement that is as friendly as possible. Then see if your design matches actual plants and pollinators in the ecosystem and see whether nature evolved something even more interesting than you came up with.

Note The models in this chapter are not intended to be precise models of any given plant. Rather, they are a demonstration of how relatively simple rules can create models that can show some broad principles of plant growth. The realities of plant evolution are complex; we suggest further reading at the end of the chapter. That said, we think that creating a plant from first principles to live in an environment is an exercise that gives some unusual insight into the problems plants (and home gardeners) deal with every day. This section shows you examples of what the code can do. Then, in the next section, we talk about how we created these and how you can create models of your own.

Desert Plants

The first plant we tried to print out was a desert plant, like an aloe or agave. Figure 6-3 shows our print next to the real thing. Desert plants tend to have their leaves or other major parts arranged so that water will run right down to their roots if it rains and thus are relatively easy to print, with limited overhangs.

Figure 6-3. *3D printed aloe model next to its real cousin*

CHAPTER 6 PLANTS AND THEIR ECOSYSTEMS

However, this model was challenging to print because the leaves became very thin at the top, and printers have difficulty printing sparse, thin material at the top of a print, so we tended to get a bit of stringing. Figure 6-4 shows the model completed, and Figure 6-5 is a close-up showing the stringing. The drooled bits can be trimmed off carefully with small, pointed scissors or another tool. Wear eye protection because the bits tend to fly around when you snip them off.

Figure 6-4. *The finished aloe*

CHAPTER 6   PLANTS AND THEIR ECOSYSTEMS

Figure 6-5. A close-up showing stringing

Tropical Jungle Plants

Plants that live in a hot, wet environment must deal with an opposite set of problems from their cousins who live in a desert. In a hot, wet environment, plants are so dense that the biggest competition is for light. Brian Capon's book *Plant Survival* (Timber Press, 1994) says that only about 5% of the light falling on the jungle canopy makes it to the floor, which means that often there are not a lot of plants on the forest or jungle floor.

Many houseplants (such as philodendrons) evolved in a tropical jungle environment, where they survived with little light and the ability to handle being overwatered. Figure 6-6 is a large philodendron, commonly known as the lacy tree.

CHAPTER 6   PLANTS AND THEIR ECOSYSTEMS

Figure 6-6. *Lacy tree (Philodendron bipinnatifidum) enjoying a damp, shady corner of a Southern California yard*

The other problem a jungle plant faces is avoiding having so much water around that the plant will rot. Some plants make a living hanging off trees (such as *epiphytes*) and get their moisture from the air. Plants develop waxy leaves and "drip tips" on the leaves to shed water, like the calla lily leaf in Figure 6-7. (Note that waxy leaves can keep water in, too, and are not limited to jungle plants.) Since many jungle plants are spindly (to get maximum light) and have big, curved leaves, they are challenging to model with a 3D print. The leaves want to be nearly parallel to the sun to capture as much light coming down as possible, though, so we can print the leaves nearly flat, as we will see in a minute.

Because of the water-shedding downward tips of a typical jungle leaf, these plants required a different model than the desert ones. This

CHAPTER 6　PLANTS AND THEIR ECOSYSTEMS

model creates the leaves one at a time and creates a base with stems that allows you to lay out the stems in a spiral pattern and print big, flat leaves separately. We talk about the models in detail (and provide listings and settings) in the "Printing the Models" section later in the chapter.

Figure 6-7. *Leaves of a plant native to a wet environment*

Figure 6-8 shows the parts of the jungle plant model, and Figure 6-9 shows it assembled. If you were creating an ecosystem, you would most likely glue the leaves in place (and probably glue the base to the bottom of any display you were creating). If you wanted the leaves to be a little more horizontal, you could accomplish that with a bit of glue.

CHAPTER 6   PLANTS AND THEIR ECOSYSTEMS

Figure 6-8. Plant with drip-tip leaves, not yet assembled

Figure 6-9. Assembled version of the model in Figure 6-8

CHAPTER 6   PLANTS AND THEIR ECOSYSTEMS

Figure 6-10 is a photo of the jungle plant model taken from above (like the photo of the real plant in Figure 6-6 or Figure 6-7). Each photo is taken with light impinging on the plant from different angles. Notice how well the leaves fill the space (and how little one leaf shades another) for these varying lighting conditions.

Figure 6-10. *The jungle plant model from above on a sunny day*

Flowers

Plants make their food through photosynthesis, using water and solar energy converted by their leaves. But they also need to keep their species going. Some plants reproduce by sending out runners or budding off pieces, cloning themselves. Others need to have pollen moved around from male flower parts to female ones. To do that, certain types of plants (*angiosperms*) have evolved flowers to attract and guide pollinators.

As with leaves, it is an advantage to a plant to use available energy to make a flower as efficiently as possible. A flower's job is to get the attention of a pollinator (like a bee) and attract it into the flower so that the pollinator gets coated with pollen, which then gets carried to another flower for pollination.

As you can see in Figures 6-11, 6-12, and 6-13, the two formal *Camellia japonica* blossoms and the lowly (some might say weedy) lawn daisy all have their petals splayed out for maximum display. We used the same mathematics to create the flowers as we did for the aloes, with some different values for variables. Details are in the "Printing the Models" section.

Figure 6-11. *A species of C. japonica, not yet fully open*

CHAPTER 6 PLANTS AND THEIR ECOSYSTEMS

Figure 6-12. *Another species of C. japonica*

Figure 6-13. *Lawn daisies*

Figure 6-14 shows the camellia model (printed without support), and Figures 6-15 and 6-16 show the daisy. You can see that this simple model replicates a lot of the structure, although obviously no model this simple will be a perfect one. In the next section, we tie together these models and talk about how to create new plants and flowers with them.

CHAPTER 6   PLANTS AND THEIR ECOSYSTEMS

Figure 6-14. *3D printed camellia model*

Note that the daisy (Figures 6-15 and 6-16) model required some support—a 3D printer builds up these models from the bottom, and petals that are nearly horizontal need to have some support material printed below them so that the petals or leaves are not printed in air. In this case, we used PrusaSlicer's organic support option and a support angle of about 35 degrees to avoid printing excessive support structures. More extensive support structures might have been difficult to break away without damaging the relatively delicate petals.

215

CHAPTER 6   PLANTS AND THEIR ECOSYSTEMS

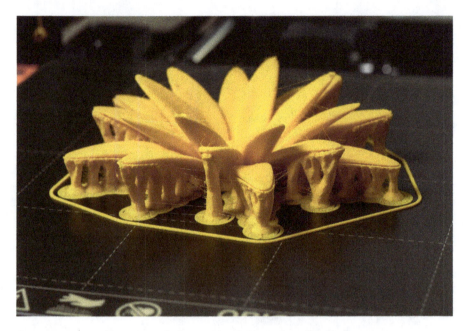

Figure 6-15. 3D printed daisy model on the platform, showing support

Figure 6-16. 3D printed daisy after support removal

Printing the Models

Two models collectively generated everything in the previous section. The first one, in Listing 6-1, created the plants with leaves attached at the plant base (like the aloe) and the flowers. Listing 6-1, with appropriate parameters, also created the base and stems for jungle plants. The leaves, petals, or stems are laid out in the spiral we described in the earlier section "The Mathematics of Plant Growth." Listing 6-2 generates one jungle plant leaf—the model had to be different since these leaves are relatively flat and horizontal, which would require a lot of support material to print on their stems.

The models assume that the leaves or petals spiral out from a center. How tightly they spiral, whether any space is left in the center, whether they tilt or get bigger as they spiral out, and similar considerations are controlled by several variables that you can tweak to create models of different plants.

Plant and Flower Models

The plant/flower model creates a petal/leaf that draws on the curved, variable-thickness model we created for airfoils in Chapter 4. Then this leaf is duplicated, and each new leaf is rotated, stretched, and offset to build up the model. In the case of the jungle plant stems, the parameters listed in Table 6-1 will give you stems, on which you can hang the leaves created in Listing 6-1. The parameters you can change in the model in Listing 6-1 are as follows:

- length: The length of each petal (mm). However, the model will stretch the petals horizontally (along the radius) while shortening them vertically, based on the value of some of the other variables.

- width: The base width of each petal (mm). Depending on other variables, this should be roughly the width of the leaf at its widest point.
- thickness: The thickness of each petal as a percentage of width.
- pointiness: A parameter that should be greater than zero but not exceed 1, defining the shape of the end of the leaf. A value of 1 will make a leaf with a pointed tip, while a value close to 0 will make a leaf that is almost circular.
- curvature: A parameter that controls how much the petals curve outward as their number increases. The range for this number will depend on how many petals you have as well as the size of the gap in the center (if any), but you are unlikely to need a value larger than about 5. If your outer petals get long and flat as they radiate horizontally, decrease this number.
- shorten: How much to squash down (in the vertical direction) the outer leaves relative to the inner ones. If things start pointing downward around the outside of the plant, shorten is too big.
- petals: Number of petals. Should be a number in the Fibonacci sequence (see discussion earlier in the chapter).
- petalSpacing: Radial offset of each petal relative to the next one (cumulative, mm).
- openness: Incremental angle (cumulative, in degrees) of how much each subsequent petal is bent down at its base relative to the previous petal.

CHAPTER 6 PLANTS AND THEIR ECOSYSTEMS

- `petalOffset`: Skip this many petals in the center, which creates an open space and a flatter flower.
- `tip`: Diameter of the leaf tip, in mm. Should usually be 0 unless you want the end of the petal to have some minimum width (such as for adding the drip leaves).

Table 6-1 gives the values of parameters for the test cases earlier in the chapter. In the model, these variables strongly interact with each other to change the geometry. Be sure you look carefully at the model visualization in OpenSCAD and your 3D printer's slicing program before committing to a print. Some combinations may create conditions that will at best be difficult to print and at worst may crash OpenSCAD. Do not give up too easily in OpenSCAD, though, because the rendering can take a while—half an hour or so—even on a modern computer. Always use OpenSCAD's preview function to quickly visualize what your model will look like before attempting the slower render function.

In addition to the OpenSCAD model, we have included 3D printable STL files `camellia.stl`, `daisy.stl`, `aloe.stl`, and `jungleStem.stl` for you to directly print. Note that to get the jungle plant, you also need to print the leaves (with `DripLeaf.stl`).

Caution Since many of the variables have open-ended values, it was not possible to exhaustively test the system. We suggest you start with one of the examples in Table 6-1 and then tweak the variables to get the plant you would like. If you get very big outer petals/leaves, decrease the curvature value. These models aim to mimic the general appearance of plants, not capture their process of adding leaves or detailed structure of petal and leaf attachment.

CHAPTER 6 PLANTS AND THEIR ECOSYSTEMS

Table 6-1. *Values of the Variables in* `FlowersPlants.scad` *for the Examples in This Chapter*

Variable	Aloe	Daisy	Camellia	Jungle Stem
`length`	50	50	30	120
`width`	6	3	3	3
`thickness`	0.3	0.3	0.1	1
`pointiness`	0.5	0.2	0.1	1
`curvature`	1	0.25	0.2	3
`shorten`	3	0.25	15	6.5
`petals`	34	21	89	8
`petalSpacing`	0.4	0.4	0.2	0.2
`openness`	0.2	0.9	0.1	0.2
`petalOffset`	10	30	21	13
`tip`	0	0	0	3

Note The daisy model as shown in Figures 6-15 and 6-16 was printed at 80% of the size the variables shown in Table 6-1 would generate. The others were printed at the size naturally generated by these values. As a rule, it is better to use the parameters to scale a model rather than scaling in the slicing program.

Note that for these four cases, you do not need to set all of these parameters individually. If you open OpenSCAD's `Customizer` (as described in Chapter 1), you will find that there are four corresponding cases already defined in the `Preset` list. Selecting one of those will automatically set all the parameters to the appropriate values for, say, an aloe.

CHAPTER 6 PLANTS AND THEIR ECOSYSTEMS

However, if you use the Customizer presets, you cannot tweak the other values. To make a variant of one of these listed, you need to set the values for the parameters listed individually and not use the pull-down menu for presets. To learn more about the Customizer, see Chapter 1.

To make a custom plant, you first need to make sure that the preset option is blank. With no preset selected, you can use the following variables to create your own plant. To do this, you probably want to use one of the sets of values in Table 6-1 as a starting point and then type in the values.

For example, for the small aloe in Figure 6-3, we did not simply scale the bigger one. We changed length to 25, thickness to 0.2, petals to 21, and minbase to 10. This gave us a plant that was a bit more tightly packed and had shorter leaves, as happens in the real plant. But all this needs to be typed in (starting with hand typing in the other aloe values that are different from the defaults).

Listing 6-1. *Flowers and Plants Growing from a Base (file FlowersPlants.scad)*

```
//OpenSCAD program to create plants and flowers in
//a modified Archimedes spiral
//(c) 2016-2024 Rich Cameron
//for the book 3D Printed Science projects, Volume 1
//Licensed under a Creative Commons, Attribution,
//CC-BY 4.0 international license, per
//https://creativecommons.org/licenses/by/4.0/
//Attribute to Rich Cameron, at
//repository github.com/whosawhatsis/3DP-Science-Projects

//presets override the settings in the next section
preset = 0; // [0:"", 1:aloe, 2:daisy, 3:camellia, 4:jungle]
/*[custom plant settings]*/
//base petal length (will be modified), mm
```

CHAPTER 6   PLANTS AND THEIR ECOSYSTEMS

```
length = 50;
//base width, mm
width = 6;
//base thickness, percentage
thickness = .3;
//should be between 0 and 1 - pointier leaves have higher value
pointiness = .5;
//makes the _petals bend more
curvature = 1.5;
//makes the petals get shorter from the center out
shorten = 6;
//needs to be a Fibonacci number
petals = 21; // [1, 2, 3, 5, 8, 13, 21, 34, 55, 89, 144]
//radial offset of each petal (cumulative)
petalSpacing = .4;
//angle of each petal (cumulative)
openness = .2;
//skip petals in center, creates open space and flatter flower
petalOffset = 15;
//diameter of tip
tip = 0;

/*[other settings]*/
//mm, increase if your base is too small
minbase = 20;
//vertical segments, more = smoother, but longer rendering
zseg = 20;
//horizontal segments, more = smoother, but longer rendering
xseg = 10;

{} // end customizer

goldenAngle = 137.508; //constant, should never change
```

```
presets = [[],
  // l,   w,  t,  p,   c,    s,   p,  pS,  o,  p0, t
  [ 50,  6, .3, .5,    1,    3,  34,  .4, .2,  10, 0],
  [ 50,  3, .3, .2,  .25,  .25,  21,  .4, .9,  30, 0],
  [ 30,  3, .1, .1,   .2,   15,  89,  .2, .1,  21, 0],
  [120,  3,  1,  1,    3,  6.5,   8,  .2, .2,  13, 3],
];

_length = preset ? presets[preset][0] : length;
_width = preset ? presets[preset][1] : width;
_thickness = preset ? presets[preset][2] : thickness;
_pointiness = preset ? presets[preset][3] : pointiness;
_curvature = preset ? presets[preset][4] : curvature;
_shorten = preset ? presets[preset][5] : shorten;
_petals = preset ? presets[preset][6] : petals;
_petalSpacing = preset ? presets[preset][7] : petalSpacing;
_openness = preset ? presets[preset][8] : openness;
_petalOffset = preset ? presets[preset][9] : petalOffset;
_tip = preset ? presets[preset][10] : tip;

function chord(z) = max(
  _tip / _width / 2,
  sqrt(z) * sqrt(1 - z) * 4 * pow(1 - z * _pointiness, 2) *
    pow(1 / _pointiness, 1/3)
);
function camber(x, a = .5, b = 1) = a * (cos(x * 180 * b) + 1)
  / 2;
function theta(x, a = .5, b = 1) = a * -sin(x * 180 * b) * 60;
function _thickness(x, a = 1, b = _thickness) = a *
  pow(x + 1, 2) * pow(x - 1, 2) / 10 +
  b * (sqrt(1 + x) + sqrt(1 - x) - 1);
```

CHAPTER 6   PLANTS AND THEIR ECOSYSTEMS

```
$fs = .5;
$fa = 2;

//create triangles from quad
function quad(a, b, c, d, r = false) = r ?
  [[a, b, c], [c, d, a]]
:
  [[c, b, a], [a, d, c]]
;

function points(roll = 2) = concat(
  [for(i = [for(z = [for(z = [0:zseg]) z / zseg])
    concat(
      [for(
        x = [for(i = [-xseg:xseg]) i / xseg],
        a = chord(z) * (x + _thickness(x) *
          sin(theta(x, .1, chord(z) / 2))),
        r = max(
          .1,
          ((camber(x, .1, chord(z) / 2) - _thickness(x)) *
            chord(z) * cos(theta(x, .1, chord(z) / 2))) +
            roll + _curvature * pow(z * roll, 2)
        )
      ) [
        (sin(min(max(a * 180 / PI / r, -179), 179))) * r,
        (cos(min(max(a * 180 / PI / r, -179), 179))) * r -
          roll,
        z
      ]],
      [for(
        x = [for(i = [-xseg:xseg]) i / xseg],
        a = chord(z) * (-x + _thickness(x) *
```

```
            sin(theta(x, .1, chord(z) / 2))),
        r = max(
          .1,
          ((camber(x, .1, chord(z) / 2) + _thickness(x)) *
            chord(z) * cos(theta(x, .1, chord(z) / 2))) +
            roll + _curvature * pow(z * roll, 2)
        )
      ) [
        (sin(min(max(a * 180 / PI / r, -179), 179))) * r,
        (cos(min(max(a * 180 / PI / r, -179), 179))) * r -
          roll,
        z
      ]]
    )
  ], j = i) j]
);

faces = concat(
  [for(i = [for(x = [0:xseg * 2], xmax = xseg * 4 + 1)
    quad(
      x,
      x + 1,
      xmax - x - 1,
      xmax - x,
      true
    )], v = i) v],
  [for(i = [for(
    z = [0:zseg - 1],
    x = [0:xseg * 4 + 1],
    xmax = xseg * 4 + 2
  )
    quad(
```

CHAPTER 6   PLANTS AND THEIR ECOSYSTEMS

```
        z * xmax + x,
        z * xmax + x + xmax,
        z * xmax + x + (((x % xmax) == xmax - 1) ? 0 : xmax) + 1,
        z * xmax + x - (((x % xmax) == xmax - 1) ? xmax : 0) + 1,
        true
      )], v = i) v],
    [for(i = [for(x = [0:xseg * 2], xmax = xseg * 4 + 1)
      quad(
        len(points()) - 1 - (x),
        len(points()) - 1 - (x + 1),
        len(points()) - 1 - (xmax - x - 1),
        len(points()) - 1 - (xmax - x),
        true
      )], v = i) v]
);

echo(presets[preset]);

for(petal = [_petalOffset:_petalOffset + _petals - 1])
  rotate([0, 0, petal * goldenAngle])
    translate([0, petal * _petalSpacing, 0])
      rotate([-petal * _openness, 0, 0])
        scale([
          _width,
          _width,
          _length *
            (1 - pow(
              petal * _openness * _curvature * _shorten / 100,
              3
            ))
        ]) polyhedron(
          points(petal * _petalSpacing / _width),
          faces
```

CHAPTER 6   PLANTS AND THEIR ECOSYSTEMS

```
      );
translate([0, 0, -_width / 2])
  scale(max(
    minbase,
    ((_petalOffset + _petals) * _petalSpacing + _width)
  ) / 50)
    intersection() {
      scale([1, 1, .5]) sphere(50);
      translate([0, 0, 50]) cube(100, center = true);
    }
```

Jungle Plant Leaf Model

Listing 6-2 generates jungle plant leaves that have a connector that hooks onto the base and stems generated by the model in Listing 6-1 (with appropriate parameters). The variables in Listing 6-2 are as follows:

- size: Length of the leaf from the drip tip to base (mm)
- hole: Size of the opening in the connector (mm)
- waviness: How wavy the outer boundary of the leaf is, from 0 (not wavy) to 1 (example shown in Figure 6-8)

Note The jungle leaves and stems in Figures 6-7 through 6-9 were printed from the models in Listing 6-1 (with the jungle parameters from Table 6-1) and eight copies of the leaf generated by Listing 6-2. We have included the STL file DripLeaf.stl; create a total of eight copies of it in your slicer program to populate the tree base created by default.

Listing 6-2. Jungle Plant Leaves (file DripLeaf.scad)

```
//OpenSCAD program to create leaves with "drip tips"
//(c) 2016-2024 Rich Cameron
//for the book 3D Printed Science projects, Volume 1
//Licensed under a Creative Commons, Attribution,
//CC-BY 4.0 international license, per
//https://creativecommons.org/licenses/by/4.0/
//Attribute to Rich Cameron, at
//repository github.com/whosawhatsis/3DP-Science-Projects

size = 50;
hole = 4;
waviness = 1;

$fs = .2;
$fa = 2;

difference() {
  union() {
    linear_extrude(.6) scale(size / 25) for(j = [0, 1])
      mirror([j, 0, 0])
        for(_i = [for(i = [0:100 - 1]) i / 100])
          hull() for(i = [_i, _i + 1/100])
            translate([0, pow(i, 2) * 10, 0])
              rotate(180 * sqrt(i))
                scale([
                  .1,
                  pow(i, .5) * 10 - (-cos(i * 10 * 360) + 1) *
                    waviness
                ]) rotate(-135) square();
    linear_extrude(1) scale(size / 25) {
      for(j = [0, 1]) mirror([j, 0, 0])
```

```
      for(i = [for(i = [0:10]) i / 10])
        translate([0, pow(i, 2) * 10, 0])
          rotate(180 * sqrt(i)) hull() {
            scale([.1, pow(i, .5) * 10]) rotate(-135)
              square();
            circle(pow((i + 1), 1) * .5);
          }
    hull() for(j = [0, 1]) mirror([j, 0, 0])
      for(i = [for(i = [0:10]) i / 10])
        translate([0, pow(i, 2) * 10, 0])
          rotate(180 * sqrt(i))
            circle(pow((i + 1), 1) * .5);
  }

  linear_extrude(hole * 2, convexity = 5) difference() {
    circle((hole / 2 + 1.2));
    circle((hole / 2));
  }
}
linear_extrude(.8, center = true)
  scale(size / 25) offset(-.025) {
    for(j = [0, 1]) mirror([j, 0, 0])
      for(i = [for(i = [0:10]) i / 10])
        translate([0, pow(i, 2) * 10, 0])
          rotate(180 * sqrt(i)) hull() {
            scale([.1, pow(i, .5) * 10]) rotate(-135)
              square();
            circle(pow((i + 1), .5) * .1);
          }
    hull() for(j = [0, 1]) mirror([j, 0, 0])
      for(i = [for(i = [0:10]) i / 10])
        translate([0, pow(i, 2) * 10, 0])
```

```
            rotate(180 * sqrt(i))
                circle(pow((i + 1), .5) * .1);
    }
}
```

Printing Suggestions

Some of the models in this chapter may require support, as mentioned. Your 3D printer's slicing program generates support automatically if you tell it to (see Chapter 1). If a petal or leaf is hanging out in the breeze and is significantly flatter than a 45-degree angle to the platform, you probably need support. The daisy was this way; the camellia was printable without support. If the contact area with the platform is small, you may also want to use a raft (a few extra layers under the print) as we describe in Chapter 1.

If you want to test whether your outer petals will need support, you can increase the petalOffset value to be equal to the value you plan to use plus two or three less than your total planned petals. For instance, we tested a three-petal camellia to see if we could print it without support. The full camellia has `petals = 89` and `petalOffset = 21`; we ran a test with `petals = 3` and `petalOffset = 21 + 89 - 3 = 107`. This allowed us to only print the outermost three petals—the ones most likely to need support—to see whether we could get away with doing the rather lengthy camellia print without support. (The answer was we could—see Figure 6-17 for the test print.) However, since this is a variation on the camellia, you will need to put in all the camellia values by hand in the program and then change a few as noted here.

CHAPTER 6　PLANTS AND THEIR ECOSYSTEMS

Figure 6-17. *Testing a "partial camellia"*

Tip　Some of these models take quite a long time to render in OpenSCAD. They may also take a while to print. Many of the models we show in this chapter each took many hours on our relatively fast 3D printers. Plan ahead if you want to use them in a class.

THINKING ABOUT THE MODELS: LEARNING LIKE A MAKER

This chapter started with the idea that it would be interesting to see how well the math described earlier in this chapter would create plants with plausible-looking (or at least plausibly functioning) structures for different environments. Fairly rapidly, though, we needed to go out and look at as many plants as we could to see how the plants handled things like overlapping leaves and petals and how real plants attached and added on structure. At some point, Rich realized that he could adapt the techniques used to describe the shape of an airfoil from Chapter 4 to create petal/leaf individual objects. That plus the basic spiral distribution got us underway.

Showing plants that evolved in wet and dry environments seemed like a good way to showcase two opposite ecosystem types. However, we realized early on that a desert plant for the most part is a natural match to 3D printing. Many

CHAPTER 6 PLANTS AND THEIR ECOSYSTEMS

desert plants have structures that point straight up or gradually bend inward from a broad base since the plants tend to want to encourage water runoff right to their small root systems. This makes them relatively easy to capture in a 3D print.

Similarly, the basic flower blooms we were modeling could be delicate prints but could be printed flat and with no or minimal support just from the platform. (There are, of course, many flower blooms that were not really feasible on a consumer-level printer, but the ones here will let you think about the issues of attracting pollinators in different environments.)

Jungle plants, with their arched leaves and suspended airy leaves, were a different story. To maintain the water-shedding arched shapes of many leaves in these environments, we printed out leaves individually and put them into a base that spaces them in an Archimedes spiral. We decided against modeling any plant leaf in detail but instead focused on the sparse foliage and water-shedding structure of the plants like calla lilies and philodendrons.

Briefly, we thought about trying to accurately model the process that plants use to add a leaf or petal. We decided in the end to just empirically create a model that gave a physically-plausible model. This got big-picture features right and was more flexible and more broadly applicable. (We did not want to lose sight of the forest for the trees, as is were.) The models are very nonlinear, and the parameters have to allow for a lot of shapes, which can result in some interesting interactions. We encourage you to play with changing these parameters, thus "evolving" some cool species of plants. Think about where these imaginary flora might flourish in the real world.

Where to Learn More

There are a lot of good botany books aimed at the public. Many are aimed at gardeners, such as Brian Capon's books *Botany for Gardeners* (Revised edition, Timber Press, 2005) and *Plant Survival: Adapting to a Hostile World* (Timber Press, 1994). As the titles would imply, the first one is a general introduction to how plants make a living, whereas the second is a great survey of how a plant and its ecosystem interact.

One book that was helpful in thinking about creating these models was Lizabeth Leech's book *Botany for Artists* (The Crowood Press, 2011), which had many wonderful structural discussions and deconstructions about how to think about and represent plant structure. If you want to do more artistically sophisticated 3D models than the ones shown in this chapter, you might find Leech's book very helpful both in how to look at and analyze a plant's structure and for the many very detailed, clear pictures and diagrams of plants.

To learn more about extinction in a rapidly changing world, Elizabeth Kolbert's Pulitzer Prize–winning book *The Sixth Extinction* (Picador, 2015) is a good nontechnical review of the issues and the people trying to understand them.

If you would like to do chemistry experiments with plants and understand the science behind their coloration, David Lee's book, *Nature's Palette: The Science of Plant Color* (University of Chicago Press, 2007), has some great experiments with basic hardware-store materials that might be interesting for a lab course.

We have given some references for the mathematics of the golden ratio and the Fibonacci sequence where we introduced the topic. The Khan Academy (www.khanacademy.org) has many great videos about the underlying math as well as where the ratios occur in nature and architecture. Search on "golden ratio" and "Fibonacci" for more.

CHAPTER 6   PLANTS AND THEIR ECOSYSTEMS

If you live near a major botanical garden or can take a virtual tour of one, nothing compares to seeing plant communities in the wild. We highly recommend seeing some full-scale, real ecosystems before trying to design your own.

Teacher Tips

As noted earlier, we could imagine these models being used just as literal plants to play with in the K–2 grades. See the K–2 science standards about interrelationships between plants, animals, and where they live, such as K-ESS3-1 at www.nextgenscience.org/topic-arrangement/kinterdependent-relationships-ecosystems-animals-plants-and-their-environment.

In middle school, standards such as MS-LS2-5 may apply—see www.nextgenscience.org/pe/ms-ls2-5-ecosystems-interactions-energy-and-dynamics.

In high school, these models might find a use when teaching standards such as HS-LS2-7, designing solutions to reduce human impacts on biodiversity (www.nextgenscience.org/pe/hs-ls2-7-ecosystems-interactions-energy-and-dynamics).

These are just a few topics that struck us as interesting tie-ins to study the concept of considering what plant features would make the plant more likely to survive in an ecosystem. We are sure that teachers will come up with better ones.

If this exercise inspires you to go out and observe plants in the wild, you can help with a citizen science project and gather data. If you browse projects on www.inaturalist.org, you can probably find a botanical project near you where you and your students can contribute observations.

CHAPTER 6   PLANTS AND THEIR ECOSYSTEMS

Science Fair Project Ideas

You can build a science fair project around these models in several directions. First, you could simply try changing the different variables in the models and seeing what sort of plants you get based on the mathematics and compare them to real plants. Do real flowers with thin, widely separated petals tend to always have 3, 5, or 8 petals, for instance? How often do plants fit the rule? Do plants of a species always have the same number of petals?

Alternatively, you could start with a design goal, like "I am going to design a garden for a very sunny, hot, dry climate." How would you shape the plants—and mix them into a community—to use the available sun, manage excess sun, and use water well? What does your community look like? Does it look like real desert plant communities? How would your plant community fare if the climate changed suddenly? Analyzing the differences between real plant communities and your constructed ones might lead you to some interesting explorations of climate, evolution, and mutations.

You might also consider what would happen if you inserted an "invasive species" into your carefully balanced plant grouping. What might it displace? What else might happen to the other organisms (bugs, microbes, and so on) in the ecosystem? Where do they live? You could also imagine creating a game that would track which plants got enough (or too much) water, sun, or nutrients to see which species might be the winners and losers in the presence of an invasive species.

Summary

In this chapter, you learned about different types of plants and how they have evolved structures that make them successful in their environments. We developed some simple ways to create notional models of plants and explored how one might use these models to think about plants and their roles in their ecosystems.

CHAPTER 7

Molecules

A lot of the art of chemistry involves being able to visualize abstract descriptions of activities that are occurring on the atomic or molecular level. Often these abstractions can be described mathematically with very complex equations that most people will not see until later years in college, if then. In this chapter, we show you how to create a few minimalist models that may help you develop intuition about these interactions.

Chemistry books are usually dense with diagrams and equations. As a conscious departure from that, we illustrate the concepts in this chapter with photographs of the models themselves so that you can see how the explanation and the model go together.

First, we give you some chemistry background and explain the terminology for our models. Then we talk about how we came up with these models and how we think you might use them to teach or learn the concepts embodied in them and perhaps how to extend the models for your own uses or science projects. Rather than create a giant set of models that captures many circumstances, we have stuck to just three: an idealized atomic structure that applies in principle to atoms of six different elements, a carbon atom (a special case of the first model), and a water molecule. Then we look at how these atoms are incorporated into crystals.

CHAPTER 7   MOLECULES

MODELS USED IN THIS CHAPTER

This chapter uses three different OpenSCAD models. For more on 3D printing and creating and using math models in OpenSCAD, check out Chapter 1. Select 3D printable STL example files are included in the repository as well. The OpenSCAD models are

- `orbitals.scad`: A model of atomic structure that applies to six different elements
- `carbon_tetrahedron.scad`: A model of atomic structure that applies to carbon atoms
- `water.scad`: A model of the structure of water molecules that can be assembled into ice crystals

To do two of the model builds described in this chapter, you will need to print a total of about 30 of the water molecules to create ice crystals, preferably 15 each of two different colors. Note that the model `carbon.scad` in the first edition of this book has been renamed `orbitals.scad` since that model applies more accurately to elements other than carbon. The new `carbon_tetrahedron.scad` model is a better representation of carbon's structure.

Chemistry Background

Atoms are made up of smaller subatomic particles called *protons* (which are positively charged), *electrons* (negatively charged), and *neutrons* (not carrying a net charge). They typically have equal numbers of protons and electrons to have no net charge, but there are exceptions to this. Atoms form chemical *bonds* with one another through an interaction between their *electrons*. In this way, molecules are created out of atoms.

CHAPTER 7 MOLECULES

Electrons are very tiny particles with a negative electric charge that whir around the positively charged, relatively large nucleus of the atom. About 100 years ago, physicist Niels Bohr proposed that the electrons orbit the nucleus in much the same way as planets orbit a star in a solar system. This is good for visualizing a very rough model of an atom, but the reality is more complicated than that and rapidly gets into the realm of quantum mechanics.

You can buy kits of varying levels of sophistication to create physical models of chemical bond structure. Most of them fall into two categories: either very simple models of the basic shapes of the "clouds" of electrons that interact with other atoms to form molecules or "ball and stick" models of how atoms go together to create molecules. Both have their good points and shortcomings, and in this chapter, we talk about our process of coming up with a few minimalist models that you can use as the basis of your own explorations.

Chemistry can seem like a maze of unconnected facts to memorize that have no guiding principles to hang them all together. However, the *periodic table of the elements* gives us some rules and organization about how atoms are constructed and how they come together with other atoms.

The Periodic Table of the Elements

The periodic table of the elements is a classic depiction showing the relationships between different atoms in a compact graphical way. A tremendous amount of information is embodied in the periodic table, and some nuances may be important if you decide to do an independent project in this area. Rather than capture a subset of it here in a small diagram, we suggest you look at one of the many online versions of the periodic table. (Just search on "periodic table.") Wikipedia's "Periodic Table" entry is particularly good, as is www.rsc.org/periodic-table (which includes podcasts and many links from Britain's Royal Society of Chemistry) and https://pubchem.ncbi.nlm.nih.gov/periodic-table/. There are more resources at the end of the chapter.

CHAPTER 7   MOLECULES

Figure 7-1 shows the broad structure of the periodic table. Each box represents an element. We will be talking about elements in the top row of the yellow part of the table for the most part in this chapter. The elements in groups (columns) 1, 2, and 13–18 of the periodic table are called the *representative elements*, or sometimes the *main group*. Figure 7-1 has these columns colored pink or yellow. There are various conventions about what to call the different colored blocks in Figure 7-1, and you will also see the table split up in ways different from that we used here. Figure 7-1's format is based on how the electrons of the atoms are arranged.

The columns are called *groups*, numbered 1 to 18 from left to right; the rows are called *periods*, starting from 1 at the top and going to 7 at the bottom (ignoring the green boxes). The atomic number of the element tells us how many protons (and electrons in an electrically neutral atom) atoms of the element have. Hydrogen is the box on the upper left of Figure 7-1 and has atomic number 1. Helium is the lone pink box at the top of group 18 and has atomic number 2. Atomic number 3 is lithium, and it is in group 1, period 2. We will get to what divisions this coloring of the periodic table represents after we get a little more background.

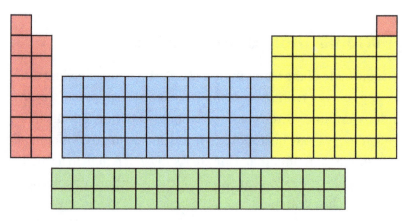

Figure 7-1. *Notional periodic table*

240

CHAPTER 7 MOLECULES

Only electrons that are the farthest from the nucleus of the atom are involved in bonding atoms to one another. These are called *valence* electrons. Atoms can share one or more valence electrons with other atoms, forming a chemical bond known as a *covalent bond*. Groups of atoms that are covalently bonded to one another are known as *molecules*. Molecules have different properties than do the individual atoms that make them up. Whether an atom will bond with another, and how it will bond, depends not just on the number of valence electrons the atom has but also on how much more room in its structure it still has left open. There are also many special cases and exceptions.

The number of valence electrons an atom has can be determined by looking at its position in the periodic table. Atoms of the elements in periodic table group 1 (the leftmost column) have one valence electron. Elements in group 2 have two valence electrons. The elements in groups 13 through 18 follow this pattern with a twist: group 13 elements have three valence electrons; group 14 elements have four valence electrons; group 15, five valence electrons; group 16, six valence electrons; group 17, seven valence electrons; and group 18, eight valence electrons.

An atom's overall number of electrons is equal to its atomic number. Things are more complicated for groups 3 through 12 (blue in Figure 7-1), and we will not address them further in this chapter. The green boxes sort of hanging off in space are another set of exceptions; they really should all be stuffed in the gap between the bottom two boxes of groups 2 and 3. The top row of green boxes are called the lanthanides, and the bottom row the actinides. We will not wade into those special cases either.

The electron structures of the atoms of the representative elements give them a great deal of flexibility in forming chemical bonds with atoms of other elements. These atoms can have at most eight valence electrons. If an atom has fewer than eight valence electrons, then it will bond with other atoms so that it can obtain a full set of eight valence electrons through sharing electrons with the other atoms. This is called the octet rule. A full octet consists of eight electrons.

CHAPTER 7 MOLECULES

Some atoms of representative elements have a full octet of eight valence electrons. These atoms (elements in Group 18) will not bond with others under any sort of normal circumstance. Thus, these elements are called the *noble gasses* since they are sort of above interacting with any other, mere peasant substance, and because they are all gasses at Earth's temperature and pressures. All the atoms of the other representative elements are lacking a full octet, so they form covalent bonds with other atoms to get their full octet. Bonding ultimately creates all the chemical compounds in existence.

Note In this chapter, we occasionally use chemical formula notation, which consists of the letters for each atom from the periodic table (e.g., C for carbon) and a subscript that says how many of that atom is in the molecule. So C_2H_6 has two carbon atoms and six hydrogen (H) ones.

Basic Orbital Shapes

Atoms are made up of clouds of electrons that take on particular shapes. The Schrödinger equation (as described in the Wikipedia article of that name) describes how electrons behave in an atom. This equation results in solutions that tell us that electrons like to fly in certain regions of space called *electron clouds*, or *probability clouds*—regions where the probability of finding an electron is high. An analogy is a swarm of gnats. Any given spot may not have a gnat in it at any time, but the swarm appears to have a shape from a distance.

Dealing with general solutions to Schrödinger's equation, though, is usually a graduate-school activity. Without going to that level, we can explore what the solutions look like and appreciate what these structures mean for the physical properties of different molecules. The clouds are

CHAPTER 7 MOLECULES

made up of orbitals—particular regions where electrons are likely to be found (see the article "Atomic orbital" in Wikipedia). As noted earlier, the atomic number of an element tells us how many electrons it will have in its orbitals. For example, hydrogen (symbol H) is atomic number 1 and has one electron. Carbon (symbol C, atomic number 6) has six.

ORBITALS

Periods (rows in the periodic table) have significance in the structure of the electron probability clouds. Each period corresponds to a shell of electrons that resides around the nucleus of the atom. The electrons of the atoms in period 1 reside in the first shell. The electrons of the atoms in period 2 reside in the first two shells and so on.

The different shapes of orbitals are denoted with letters: s, p, d, and f, in order out from the nucleus. Each orbital is made up of one or more suborbitals, and each suborbital can have up to two electrons.

The number of electrons that each orbital can hold is given by a rule called the Aufbau Principle: two electrons in the s orbital, six electrons in the p orbital, ten electrons in the d orbital, and 14 electrons in the f orbital. Each of the orbitals higher than the s orbital can have multiple suborbitals: p has three suborbitals, d has five suborbitals, and f has seven suborbitals.

The s and p orbitals are most involved in the bonding of atoms in element groups 1, 2, and 13–18. In fact, the elements in groups 1 and 2 are called the s-block of elements, and elements in groups 13–18 the p-block of elements.

In the hydrogen atom, the electrons closest to the nucleus live in a *1s* orbital. These orbitals are spherical volumes, and the electron might be anyplace in it. Depending on how many electrons the atom has, there may be multiple s orbitals, called *1s*, *2s*, and so on, up to the number of the period in which the element is found. They have the same center, but the *2s* is a larger diameter. The

CHAPTER 7   MOLECULES

1s and *2s* orbitals between them can fit the first four electrons, and so elements with atomic number 1 through 4 (hydrogen, helium, lithium and beryllium) only have *s* orbitals. Going back to the periodic table, the atoms in pink have their outermost electrons in *s* orbitals; the ones in yellow, in *p* orbitals; in blue, *d* orbitals, and in green, *f* orbitals.

The p orbitals are made up of three lobes, each of which can hold two electrons. These orbitals are sort of shaped like dumbbells at right angles to each other, extending out beyond the corresponding *s* orbital. We note what period the orbitals correspond to by adding a number before the letter. Thus, a *2p* orbital is the outermost orbital of the top (period 2) row of yellow boxes. Those atoms will have a *1s* orbital, a *2s* orbital, and three *2p* orbitals, as shown in Figure 7-2.

The model `orbitals.scad` creates a set of *1s*, *2s*, and *2p* orbitals (Figure 7-2). In principle, this structure represents the atoms with atomic numbers 5 through 10—boron (B), carbon (C), nitrogen (N), oxygen (O), fluorine (F), and neon (Ne). As we will see later, in practice, things can be a little different. Elements with more electrons will add on additional outer shells of *s* and *p* (and *d* and *f*) electrons, but we will stick to just this much for this chapter.

Figure 7-2. *The s and p orbital model*

CHAPTER 7 MOLECULES

This molecular structure consists of two *s* orbitals. One is a central ball (the *1s* orbital), and the *2s* is printed in two halves that snap together. There are three *2p* orbital pieces, as you can see in Figure 7-3. The three *p* orbitals intersect the *s* orbitals and are perpendicular to one another.

Chemists label p orbitals as p_x, p_y, and p_z, where *x*, *y*, and *z* stand for the axes in a 3D coordinate system. Since the shapes are identical and only different in orientation, we have created just one lobe model that you can place in the three orientations. In Figure 7-3, we show the parts of the 3D printed model spread out. (To avoid nasty support in the 3D print, we have designed the models in this chapter for assembly from easier-to-print parts.)

Figure 7-3. *Pieces of the model shown assembled in Figure 7-3*

Each of the orbitals, *1s*, *2s*, $2p_x$, $2p_y$, and $2p_z$, can hold a maximum of two electrons. In *quantum mechanics* (the study of how things at this scale behave), each electron is designated by a set of numbers called *quantum numbers,* sort of an identifier for that electron in that atom. Any given electron in an atom must have a unique set of quantum numbers. Two electrons that are in the same orbital, say, the $2p_x$ orbital, differ only in

CHAPTER 7   MOLECULES

their spin (called by convention *spin up* or *spin down,* an abstract quantity that has nothing to do with direction of gravity). Electrons do not spin per se, either.

The model in Listing 7-1 prints out all the pieces at once (although we chose to just print some of the pieces at a time by removing parts in the slicing program). Listing 7-1 creates models for the nucleus, s orbital halves, and p orbitals. If you wanted to print out the molecule in more than one color (as we show here), you could print two in different colors and mix and match (probably the easiest approach). Comments in the model suggest how to make these alterations.

Note   The orbitals in the model in Listing 7-1 are based on best fits to the geometry of the orbitals with some compromises for 3D printing—p orbitals do not have hooks around the s orbital, for instance, and the s orbital does not have well-defined holes in it for the p orbitals. They are not fundamental solutions of the wave equations that underlie them. In other words, these are designed to "look right" rather than derived from physics.

Listing 7-1. Molecular Orbitals Model (file `orbitals.scad`)

```
//OpenSCAD model of an atom - nucleus and s and p orbitals
//File orbitals.scad
//(c) 2016-2024 Rich Cameron
//for the book 3D Printed Science projects, Volume 1
//Licensed under a Creative Commons, Attribution,
//CC-BY 4.0 international license, per
//https://creativecommons.org/licenses/by/4.0/
//Attribute to Rich Cameron, at
//repository github.com/whosawhatsis/3DP-Science-Projects
```

CHAPTER 7 MOLECULES

```
//If you want to make one of the pieces a different color,
//you can uncomment one of the following three lines to
//make an STL of only one part, instead of the whole set.
// !s_orbital();
// !p_orbital();
// !nucleus();
//Plate the p orbitals, s orbitals, and nucleus for printing
translate([-41, 0, 0]) {
  for(i = [0:2]) translate([0, i * 22, 5]) rotate([0, 90, ])
    rotate(-16) p_orbital();
  for(i = [-1, 1]) translate([i * 20, -30, 17]) s_orbital();
  translate([20, -30, 10 * (sin(52.5))]) nucleus();
}

$fs = .2;
$fa = 2;

//Display the parts to show how they are assembled
//% modifier means they will not be included when rendering
%translate([41, 0, 0]) {
  nucleus();
  for(l = [0, 1]) rotate([180 * l, 0, 0]) s_orbital();
  for(i = [0:2]) rotate([
    i ? 90 : 0,
    i ? ((i == 1) ? 45 : 135) : 0,
    i ? ((i == 1) ? 0 : -90) : 45
  ])
    p_orbital();
}
```

247

CHAPTER 7   MOLECULES

```
//create the nucleus (sphere with a flat side for printing)
module nucleus() scale(1.0) difference() {
  sphere(10);
  translate([0, 0, -10 * (1 + sin(52.5))])
    cube(20, center = true);
}

//create s orbital halves
module s_orbital() difference() {
  intersection() {
    sphere(22);
    cube(17 * 2, center = true);
  }
  sphere(18);
  for(i = [-1:1]) rotate([0, 90 * i, 45 * i + 45]) {
    cylinder(r = 10.5, h = 100, center = true);
  }
  translate([0, 0, 50 + 2]) cube(100, center = true);
  intersection() {
    union() {
      rotate([90, 0, 45])
        linear_extrude(100, center = true, convexity = 5)
          for(m = [0, 1]) mirror([m, 0, 0])
            translate([20, 0, 0]) rotate(-5)
              translate([0, -3, 0]) square(10);
      rotate([90, 0, -45])
        linear_extrude(100, center = true, convexity = 5)
          for(m = [0, 1]) mirror([m, 0, 0])
            translate([20, 0, 0]) rotate(5 + 180)
              translate([0, -3, 0]) square(10);
    }
```

```
    translate([0, 0, 50 - 2]) cube(100, center = true);
  }
}
//create p orbital lobes
module p_orbital() difference() {
  union() {
    for(i = [1, -1]) hull() {
      sphere(2);
      translate([0, 0, i * 30]) sphere(10);
    }
    intersection() {
      sphere(12);
      linear_extrude(height = 100, center = true) hull()
        for(j = [0, 1]) translate([0, j * 15, 0])
          circle(5 - 4 * j);
    }
  }
  for(i = [1, -1]) rotate([-90, 0, i * 45]) hull() {
    sphere(1);
    translate([0, 0, 30]) sphere(10);
  }
  sphere(10);
  rotate(16) translate([55, 0, 0]) cube(100, center = true);
}
```

CHAPTER 7   MOLECULES

Assembling the Model

These models can be a little tricky to assemble the first time. First, take the nucleus and wrap one of the *p* orbitals around it. The p orbitals are designed so that their connectors will not cross each other on the nucleus. Start building the model by clipping one *p* lobe around the central (*1s*) orbital (Figure 7-4). Then take the next two *p* orbital sets and arrange them as shown in Figures 7-5 and 7-6.

Figure 7-4. *Adding the first p orbital*

Figure 7-5. *Adding the second p orbital*

CHAPTER 7   MOLECULES

Figure 7-6. *Adding the third p orbital*

Next, take the *2s* orbital halves and arrange the *p* orbitals so that they are sticking out the holes in the s orbital halves. You will see that there are slight indentations on the outside of two of the arms of the s orbital halves and on the inside of the other two. Line up the halves so that the arms that have indentations on the outside hook into arms that have them on the inside. Squeeze the arms that will be inside a little to make the model pop together. You can see the final version back in Figure 7-2.

The Carbon Atom: Hybridized Orbitals

The orbitals described in the preceding section are a fleeting thing if other atoms are around. When atoms get together, attractive and repulsive forces among the electron clouds result in a new steady state. Some shells may merge in a process called *hybridization*. This process affects the geometry of the resulting molecules. That geometry in turn largely determines the chemical properties of the molecule.

Hybridization occurs only with orbitals that contain valence electrons. The *2s* orbital of an atom combines with one or more of the *2p* orbitals to form a new orbital structure, a hybridized orbital. After hybridization,

CHAPTER 7 MOLECULES

the hybridized orbitals of the atom will contain at least one electron but at most two electrons. Each of the hybridized orbitals that contain one electron can share that one electron with a similar hybridized orbital containing one electron in another atom. Thus, the two atoms can share two electrons between them, thereby forming a covalent bond.

There can be three types of hybrid orbital: sp^1 (usually written just sp), sp^2, and sp^3. The superscript denotes how many p orbitals have been combined with the corresponding s orbital. The discussion that follows applies to carbon atoms and their compounds. Hybridization of other atoms, like nitrogen and oxygen, are a little different. We see what happens in water ice (oxygen and hydrogen) later.

An atom that is sp^3-hybridized has combined its $2s$ orbital and all three $2p$ orbitals to make four sp^3-hybridized orbitals, each one with one large lobe and one small one. An angle of 109.5 degrees exists between each of the lobes in an sp^3 hybridization, creating tetrahedral structures (with exceptions, as we will see when we talk about water). To create a methane molecule, for example, a single electron in each lobe can form a single bond with a hydrogen atom.

This type of bond between hybridized orbitals is called a sigma (σ) bond. It is formed between the orbitals of two atoms along a straight line that joins the nuclei of the two atoms. The single electrons that are in the small lobes of the sp-hybridized orbitals are not between the two atoms but are on the outside, where they can bond with other atoms. This is the type of bond that appears in diamond and water ice crystals, which we talk about in the next section.

A second type of hybridization is called *sp^2 hybridization*. An atom that is sp^2-hybridized has combined its s orbital with two of its p orbitals. Ethene (C_2H_4) is an example of a molecule in which a carbon atom is sp^2-hybridized.

The final type of hybridization is called sp^3 hybridization. In sp^3 hybridization, one of the p orbitals combines with the s orbital to make two sp-hybridized orbitals, while leaving the two remaining p orbitals unchanged. The sp-hybridized orbitals align on the same axis as each other, facing opposite directions. Ethyne (C_2H_2) has two sp^3-hybridized carbon atoms bonded with one another.

Carbon Atom Model

We have created a 3D printable model of a carbon atom: one *1s* orbital and four *2sp³* hybridized orbitals (Figure 7-7). We printed each lobe in a different color to make it a little clearer how the parts fit (Figure 7-8). Note that each lobe has one large side and one smaller one, showing the hybridized orbitals. The *1s* orbital still exists in the center, surrounded by the four *2sp³* orbitals. The model is generated by the code in Listing 7-2. The parameter size in the model can be changed to make a larger or smaller model, scaled so that features do not get too small to print.

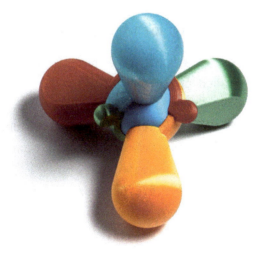

Figure 7-7. *Carbon atom model*

CHAPTER 7   MOLECULES

Figure 7-8. *Carbon atom model pieces*

Listing 7-2. Carbon Atom (file carbon_tetrahedron.scad)

```
//OpenSCAD model of a carbon atom - nucleus and sp3 orbitals
//File carbon_tetrahedron.scad
//(c) 2016-2024 Rich Cameron
//for the book 3D Printed Science projects, Volume 1
//Licensed under a Creative Commons, Attribution,
//CC-BY 4.0 international license, per
//https://creativecommons.org/licenses/by/4.0/
//Attribute to Rich Cameron, at
//repository github.com/whosawhatsis/3DP-Science-Projects

size = 20;
thick = size / 10;

$fs = .2;
$fa = .2;
```

CHAPTER 7   MOLECULES

```
translate([0, -size * 2 / 3, 0]) intersection() {
  translate(size / 2 * [0, 0, sin(55)]) sphere(size / 2);
  translate(size / 2 * [0, 0, 1]) cube(size, center = true);
}
for(i = [0:3]) translate([0, i * (size + 1), 0])
  translate([0, 0, size / 5]) rotate([0, 82, 0]) for(a = [0])
    rotate([a * (acos(1/3)), a * 180, 0]) difference() {
      difference() {
        union() {
          intersection() {
            sphere(size / 2 + thick);
            translate([-size / 10, 0, -500]) cube(1000);
          }
          translate([0, 0, -size / 5]) rotate_extrude()
            intersection() {
              translate([0, -1000, 0]) square(2000);
              union() for(lobe = size * [-.5, 1.5]) hull() {
                square(.1, center = true);
                translate([0, lobe, 0])
                  circle(abs(lobe) / 3);
              }
            }
        }
        rotate([180 - acos(1/3), 0, 0]) {
          intersection() {
            sphere(size / 2 + thick + .1);
            translate([-1000, -1000, -500]) cube(1000);
          }
          intersection() {
            translate([0, 0, -size / 5]) rotate_extrude()
              intersection() {
```

```
                translate([0, -1000, 0]) square(2000);
                union() for(lobe = size * [-.5]) hull() {
                  square(.1, center = true);
                  translate([0, lobe, 0])
                    circle(abs(lobe) / 3);
                }
              }
              translate([-1000, -500, -500]) cube(1000);
          }
        }
      }
      sphere(size / 2);
      rotate([0, -82, 0]) translate([0, 0, -1000 - size / 5])
        cube(2000, center = true);
    }
```

Assembling the Carbon Atom

The carbon atom is assembled in a similar way as was the *s* and *p* orbital model in the previous section. However, now there is no separate *2s* orbital, and the *sp* hybrid orbitals are asymmetric. Start off by adding the first hybrid orbital to the *1s* shell (Figure 7-9).

Figure 7-9. *Adding the first hybrid 2sp³ orbital to the 1s shell*

CHAPTER 7   MOLECULES

Next, add the second *sp³* orbital (red, in Figure 7-10). There is a small notch in each hybrid orbital so that the parts will lock together around the s orbital.

Figure 7-10. *Adding the second hybrid 2sp³ orbital*

Add the third hybrid orbital (Figure 7-11, green) so that it does not cross the previous two (orange and red) orbitals.

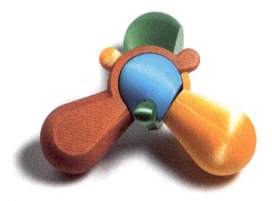

Figure 7-11. *Adding the third hybrid 2sp³ orbital*

Finally, add the fourth hybrid orbital (Figure 7-12, blue) so that it does not cross the previous two (orange and red) orbitals and hooks onto the green previously installed piece the same way the orange one hooked onto the red one. To make it fit, a bit of adjusting is required, and the tetrahedral structure results.

Figure 7-12. *Completed molecule (same image as Figure 7-7)*

Water Molecules

A water molecule consists of two hydrogen atoms bonded to one oxygen atom. The hydrogen atom bonds to one oxygen atom covalently, but hydrogen also wants to bond with nearby other oxygen atoms too. This bond between the hydrogen atom in one water molecule and an oxygen atom in another water molecule is a *hydrogen bond*. Water has some weird properties, chemically speaking. Its boiling point is much higher than one would expect, and it is denser as a liquid than it is as a solid (ice). These properties are thought to come from these hydrogen bonds.

Oxygen has six valence electrons. In a water molecule, two hydrogen atoms each bond with one of the oxygen atoms' valence electrons. This leaves two pairs of electrons on the oxygen molecule. The two hydrogens are sort of squished over to one side of the atom by these free pairs so that the atom is close to being a tetrahedron. However, it is not a regular tetrahedron. A water molecule's hydrogens are not quite at the vertices of a tetrahedron; they are slightly closer together (about 104.5 degrees, versus 109.5 degrees for a tetrahedron).

The Water Molecule Model

We have modeled one water molecule as an oxygen atom with two attached hydrogens and two "holes" (the other oxygen electron pairs), which the hydrogens of other molecules can then connect to. For 3D printing convenience, we have printed our molecules in halves, with a flat side (Figure 7-13). If you wanted just one water molecule, you could glue these together. But these were really designed to put into models of solid water—the crystalline structure we know as ice.

We have made the hydrogen bond that wants to connect to another oxygen a pair of thin plug connectors so that they will flex a bit when we create a structure from them. Listing 7-3 is the OpenSCAD model for these molecules.

CHAPTER 7   MOLECULES

Figure 7-13. *Our water molecule model (printed in two halves)*

Listing 7-3. The Water Molecule (file `water.scad`)

```
//OpenSCAD model of a water molecule for building ice crystals
//File water.scad
//(c) 2016-2024 Rich Cameron
//for the book 3D Printed Science projects, Volume 1
//Licensed under a Creative Commons, Attribution,
//CC-BY 4.0 international license, per
//https://creativecommons.org/licenses/by/4.0/
//Attribute to Rich Cameron, at
//repository github.com/whosawhatsis/3DP-Science-Projects

//Oxygen atom diameter, mm
Od = 25;
//Hydrogen atom diameter, mm
Hd = 12.5;
//Offset between centers of O and H atoms, mm
OHspacing = 12.5;
//diameter of the peg, mm
peg = 6.25;
```

CHAPTER 7 MOLECULES

```
//Tolerance (empty space) between peg in hole and peg
tol = .2;
//length of the peg
peg_len = 12.5;

$fs = .2;
$fa = 2;

angle = 104.5; //[104.5:water, 109.5:tetrahedral]

for(a = [0, 180]) rotate(a)
  translate([-Od / 3, peg_len * 1.5, 0,])
    rotate(angle / 2 - 90) half();

//Now create the half a molecule
module half() difference() {
  union() {
    sphere(Od / 2);
    for(i = [-1, 1]) rotate(angle / 2 * i) {
      translate([OHspacing, 0, 0]) sphere(Hd / 2);
      rotate([0, -90, 0])
        translate([0, 0, -(Od + Hd) / 2 - peg_len]) peg();
    }
  }
  for(i = [-1, 1]) rotate([0, angle / 2 * i - 90, 0])
    translate([0, 0, peg / 2]) peg(true);
  translate([0, 0, -100 + tol / 2]) cube(200, center = true);
}

module peg(hole = false) difference() {
  union() {
    rotate_extrude() intersection() {
```

CHAPTER 7 MOLECULES

```
      offset(hole ? tol : 0) {
        hull() {
          translate([0, peg / 2, 0]) circle(peg / 2);
          translate([0, peg_len, 0]) square(peg / 2);
        }
        if(hole) translate([peg / 2 - .25, 0d / 2 - peg, 0])
          circle(1);
      }
      translate([0, -peg, 0]) square(1000);
    }
    if(!hole) translate([0, peg / 2 - .25, 0d / 2 - peg])
      sphere(1);
  }
  if(!hole) rotate([0, -90, 0])
    linear_extrude(peg, center = true)
      offset(.45) offset(-.45) difference() {
        offset(1.5) offset(-1) square(0d / 2 - peg / 2);
        offset(.5) offset(-1) square(0d / 2 - peg / 2);
        square(peg, center = true);
      }
}
```

The Carbon Versus Water Molecule Model

In the case of the *s* and *p* orbital model and the carbon atom model, we showed lobes that "stuck out" to represent their orbitals. These models, obviously, cannot then be used to construct other models by putting together these atoms since there are no ways to hook them together. These models were purely an exercise in showing the orbitals. The water molecule model, however, has two plugs that stick out (showing the attached hydrogen atoms) and two holes that can be filled by those plugs.

CHAPTER 7   MOLECULES

Really, both carbon atoms and the water molecules like to make four connections. However, this is tricky to make in a replicable model. The water molecule is made this way to point out the two types of hydrogen bond it can make, but you should keep in mind that it is an abstraction.

Crystals

When many substances become solid, they form a crystal—a regular, repeating pattern of molecules. In this section, we look at two crystals—two kinds of water ice—and talk about similarities to diamond, which is a crystal form of carbon.

Crystals are made up of *base cells*—sets of atoms in a particular pattern that repeat over and over—in a structure called a *lattice*. For 3D printing, we find the concept of a base cell less than useful since the cells are designed to be grasped from a 2D drawing. When you can work freely in 3D, other structures are more natural, and we discuss those emergent structures here.

The water molecule model in Listing 7-3 creates just one "pair of halves" of a water molecule. Now we will take many these molecules and see what kind of structures we can create. We printed 15 of each of two colors (30 total molecules, or, if you prefer, 60 half molecules) to make our "ice crystals." A run of 15 molecules fit comfortably on a Prusa Mk3s+ print bed.

Water Ice

When water is above its freezing temperature, it flows freely. Water molecules transiently "stick to" each other as the hydrogen electrons (the plugs, in our models) come under the influence of the free pairs of electrons on other oxygen atoms (the holes, in our models). However, when the water is cooled enough, order begins to evolve from this chaos, and a crystalline form—ice—begins to form.

CHAPTER 7　MOLECULES

Depending on the circumstances, water ice can form in one of several different crystal structures; there is a good review in the Wikipedia article, "Ice." In this section, we talk about two lattice structures: ice 1h and ice 1c. Both are formed fundamentally of tetrahedra (they use the sp3 bonding covered in the sidebar about hybridization earlier in the chapter). But you will see that small differences in how the molecules connect can lead to some big structural differences.

Ice 1h

The commonest ice structure at temperatures and pressures normally encountered on the surface of the Earth is called *ice 1h*. It consists of layers of hexagons, with the layers regularly lined up one under the other. This is shown in Figures 7-14 and 7-15. (Top and side are arbitrary here, meant to show views taken at right angles from each other.)

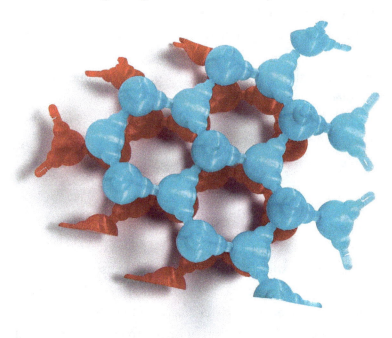

Figure 7-14. Ice 1h (hexagonal) model, bottom view

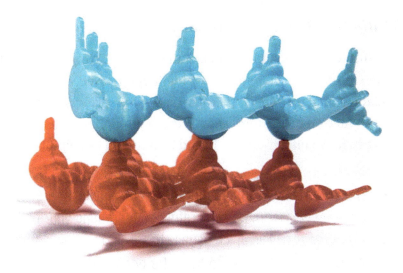

Figure 7-15. *Ice 1h model, side view*

Note You can see from Figures 7-14 and 7-15 that the water molecules in ice are quite sparsely packed into the space. This is the reason that ice is less dense than liquid water. If that were not true, then ice would sink to the bottom of lakes in winter and keep refreezing at the top until the whole lake was frozen. This would be very bad for any fish in a lake that freezes over in winter, since the layer of unfrozen water at the bottom of a frozen lake is a habitat for many species, entirely dependent on this property. Ice insulates the unfrozen water, separating it from the colder air and allowing some of the water to stay unfrozen.

CHAPTER 7   MOLECULES

Assembling an Ice 1h Molecule

Assembling the models is a little trickier than it looks. We recommend that you create the pattern apparent in the blue "bottom layer" of Figure 7-14 first and then build up from it. Alternate having the spikes stick up (and thus have a hole on the bottom) and having them stick out sideways to connect two molecules. Figures 7-14 and 7-15 should be helpful.

Note   The structure will go together in various ways that are "legal" (a peg in a hole), so you must be careful about creating your pattern consistently. This is physical; ice usually has irregularities in its crystal structure. We are showing two different ideal structures in this chapter. Joan found it easiest to think about the molecules in a layer having the bond alternately sticking up and sideways from one model to the next.

There is no particular reason why we have alternating layers of different colors—we just wanted you to be able to see how they go together. We have "half molecules" holding the molecules along the edges together, since the two halves of each water molecule do not otherwise stick. You could glue the halves together ahead of time, but that is not necessary. The "half molecules" help show how the structure would continue anyway. You can print as many molecules as will fit on your print bed by duplicating them in your slicing program.

You will probably discover that you need to hold the two halves together with one hand and the rest of the structure with the other. You might want to have the two views of the model in front of you as you lay out the first layer.

CHAPTER 7 MOLECULES

> **Tip** If you print these molecules in PLA, you may want to avoid leaving them lying about assembled, particularly in a warm environment like a parked car. There is a bit of spring force in the parts holding them together, but PLA will creep to relieve this stress, particularly in a warm environment.

Ice 1c

Another form of water ice, *ice 1c* (for more, see the Wikipedia article "Ice 1c"), is a body-centered cubic crystal lattice structure. Compared to ice 1h, it is a differently arranged repeating set of tetrahedrons. Each row is offset from the next, as shown in the top-down view of Figure 7-16 (equivalent to the view in Figure 7-14 for ice 1h). If we had a third layer on top of the white one, it would line up with the blue one and so on. On Earth, ice 1c is found in high clouds (like cirrus clouds) where the water is supercooled before ice crystals form—that is, the water in the cloud is below the freezing point for that pressure and temperature.

Figures 7-16 and 7-17 show this structure from the side and the top, equivalent to Figures 7-14 and 7-15 for ice 1h. You can see that now the hexagons are offset from layer to layer versus lying directly above one another, as they did in ice 1h (Figure 7-14).

CHAPTER 7   MOLECULES

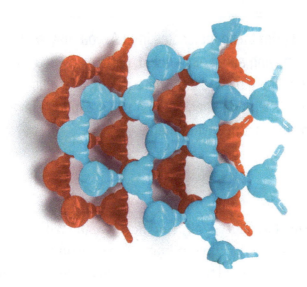

Figure 7-16. Ice 1c model, bottom view

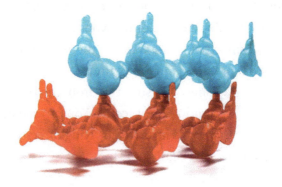

Figure 7-17. Ice 1c model, side view

CHAPTER 7   MOLECULES

Assembling the Ice 1c Molecule

Assembling the ice 1c molecule is similar to the ice 1h. Take a careful look at the two pictures of the structure and build the bottom layer, bearing in mind how the molecules alternate in their up/down patterns. The ice 1c model was built out from a central first hexagon as a somewhat more natural way to lay out this structure versus laying out several rows, as was done with the ice 1h.

Other Water Ice Structures

Water ice forms other crystal structures at low temperatures and high pressures that are created in a lab or possibly are found on other planets. These are described with numbers (ices II–XII), and most of them are not very stable. Fortunately, Kurt Vonnegut's incredibly destructive *ice-nine* (in his novel *Cat's Cradle*) is fictional. The late Barclay Kamb at Caltech did a lot of interesting work in this area; if you want to learn more, you might start by looking up some of his papers in library databases that carry scientific journals or at https://scholar.google.com. Or you can search on each of the types, like the very unstable *ice IV.*

Diamond

As it happens, ice 1c and diamonds (crystalline carbon) have very similar structures. A carbon atom can bond to four other carbon atoms in a tetrahedron, as occurs in diamond. A carbon atom similarly forms bonds to four hydrogen atoms to create methane.

If you ignore the "hydrogen atom" bumps on our ice 1c structure and the fact that the angle between the plugs there is 104.5 degrees instead of the true tetrahedral 109 degrees, the ice 1c model is a pretty good representation of diamond. Diamonds are so incredibly strong in part because each carbon atom is connected to four other ones in roughly the same configuration as ice 1c.

269

CHAPTER 7 MOLECULES

If the carbon atoms instead were strongly bonded to three others in a flat sheet and then weakly bonded between sheets, we would instead get graphite (the "lead" in pencils). If you had a structure like just one layer of ice 1c (or, for that matter, ice 1h since the single layers are the same) and allowed those to be loosely connected, you get the structure of graphite. Graphite is conductive because it has those extra electrons only loosely bonded and is weak because these layers can slide over each other.

THINKING ABOUT THESE MODELS: LEARNING LIKE A MAKER

Neither of us had taken chemistry for quite some time and so essentially had to relearn the material to create these models when we wrote the first edition. When we started researching it in earnest, we were struck by how most texts and lectures spent a very long time on nomenclature and orbitals that are not present in most compounds but are precursors to ones that are.

However, we were overwhelmed in structural detail and tried to create a general set of models that would work for a variety of molecules. Nothing clicked until we reminded ourselves of the maker dictum: find a moderately complex example, build it, and pick up the basics as you go. Then, use that as a base to learn more complex cases.

Our first attempt was to model a methane molecule: one carbon atom and four hydrogens. We came to it by sketching, using all four of our hands to hold things in place, and pestering chemist friends for physical details. At the end, we had a pretty good understanding of the chemistry and felt like we validated learning by making in the process. But then we decided that even this simple carbon model was too sophisticated relative to the level of the rest of the models in the book, and we decided to back off to just the few interesting lattices we ultimately included.

Physical model design requires some compromises. Since we are modeling something that is not a physical solid but rather clouds of electrons, of necessity, there are some constraints required to make it possible to have the models work mechanically. We have tried to minimize those constraints but could not eliminate them. To make the models easier to 3D print, parts have a strategic flat spot or two that make the models a little different than common rounded representations of these electron clouds.

We found that the constraints of making something easy to print and easy to assemble and accurate were overwhelming if one wanted to allow the user to "completely disassemble" a model of a substance like diamond or ice down to its electron orbitals. Thus, we came to the whole-molecule-level crystal element models that appear in the chapter.

In this second edition, we decided to more explicitly model the hybrid orbitals of carbon atoms, in addition to the generic *1s/2s/2p* structure we created in the first edition. If you are coming from our first edition, you will notice there are now three models in the chapter instead of two.

Joan and Rich would like to thank Michael Cheverie, a teacher of the visually impaired in Los Angeles, who initially suggested using 3D printed models to teach organic chemistry to blind students. Mike then worked with us during the writing of the first edition of this book. He helped define the carbon models and contributed to some of the chemistry background and teacher exercises that we have incorporated into this chapter and was also willing to read this chapter for the second edition as well. Glaciologist Frank Carsey was kind enough to read our discussion of ice in the first edition of this book.

Printing Suggestions

The models in this chapter are a little trickier to print and assemble than some of those in earlier chapters. Here are some rules of thumb:

- Do not scale these models up or down (particularly not down). They were created with wall thicknesses that were designed to be a little springy (which will not work if they get much thicker) or with parts that will get too thin and small to print reliably if you make them smaller. You may need to change the models a bit if you want a bigger or smaller one.

- Treat the water molecule halves as science learning tools and not general construction toys—the parts are small and could be younger-sibling choking hazards.

Where to Learn More

There are many books and websites devoted to explaining the topics in this chapter. During the development of the first edition, we referred often to Gary L. Miessler and Donald A. Tarr's text *Organic Chemistry, 2nd Edition* (Prentice Hall, 2000), to remind us of all that general chemistry background we learned once and had since forgotten.

We also enjoyed dipping around in Mariana Gosnell's book *Ice: The Nature, the History, and the Uses of an Astonishing Substance* (Alfred A. Knopf, 2005). Our companion Apress book, *3D Printed Science Projects: Volume 2* (2018) has a "Snow and Ice" chapter that continues the discussion of water ice (and adds some more models).

To learn about diamonds, besides the results of online searches on *diamond crystal structure*, we used Eric Bruton's *Diamonds, 2nd Edition* (Chilton Book Company, 1978).

CHAPTER 7   MOLECULES

Finally, here are a few online resources, with a particular focus on the orbitals section of this chapter:

- Check out the Khan Academy's extensive chemistry resources, particularly their unit "Chemical bonding and molecular structure."

- This site has excellent pictures and explanations: http://chemwiki.ucdavis.edu/Theoretical_Chemistry/Chemical_Bonding/Valence_Bond_Theory.

- The International Union of Pure and Applied Chemistry (IUPAC) officially maintains the standards for the periodic table. You can see their version here: https://iupac.org/what-we-do/periodic-table-of-elements/.

- This video explains hybridization in terms of energy: www.youtube.com/watch?v=HKyobMewXBw.

- These videos use balloons to model bonding—it becomes clear quickly why 3D printing is a bit more manageable way to go, but the visualization is fun: www.youtube.com/watch?v=bOKvfvJi-vk and www.youtube.com/watch?v=KbOmxAMHnfE.

- Chem LibreTexts (https://chem.libretexts.org/) has some good materials. Search there on "Atomic Structure Orbitals."

- To learn about water and ice, we used the Wikipedia pages linked in those sections, as well as the results of and general searches online for ice cubic lattice and ice hexagonal.

CHAPTER 7   MOLECULES

Teacher Tips

We have looked at the Next Generation Science Standards to see where the models in this chapter might most usefully fit into a curriculum. You should look at your own state and school's requirements and come up with your own best alignments. It seemed to us that the best fit might be within units covering these standards:

- HS-PS1 Chemical Reactions: www.nextgenscience.org/topic-arrangement/hschemical-reactions
- PS-1 Matter and its Interactions: www.nextgenscience.org/dci-arrangement/ms-ps1-matter-and-its-interactions

Science Fair Project Ideas

These models are useful as is to learn chemistry and to think about chemical bonding. Thus, the first "experiment" to do with them is simply to print them out and see how they fit together. Some further explorations might include building more complex molecules that also have basic close-to-tetrahedral structures. Since these are plastic models of phenomena driven by quantum mechanics, be cautious about extrapolating too far from the shapes we have given you. However, if you follow our process and think about what you are trying to model, you can get some deep insights from just trying to get these little models to work.

Summary

This chapter introduces the structure of molecules and the basics of electron orbitals. It describes some of the chemistry and structure of crystals like ice and diamond. We developed 3D printed models of molecules with *1s*, *2s*, and *2p* orbitals, as well as a model of the special case of a carbon atom with hybridized bonds. We created a model of water molecules and used it to build lattices for two kinds of ice. Finally, we gave you some ideas about how to use these as a teacher and as a basis for 3D printed science fair projects.

CHAPTER 8

Trusses

This chapter explores simple structures that make up a lot of the infrastructure around us. A *truss* is a structural element that uses the strength of a triangular structure to carry loads with relatively little structural material. Trusses are used just about everywhere that a heavy load has to be carried—bridges, roof supports, and the like.

Planar trusses are thought of as existing in two dimensions (ignoring the thickness of the components), but a space truss carries the load in three dimensions. This chapter shows you how to create a model of a planar truss. We have also included a model of a simple tensegrity structure. These structures are a special type of 3D truss, constructed from a mix of stiff and flexible elements, often with cool-looking results.

There are a lot of computer-aided design (CAD) programs around that can help you figure out the forces accurately if you build a truss in CAD. However, we think that it is also important to build a bit of engineering intuition with models. At the end of the chapter, we talk about some complementary ways to simulate some of these ideas. The models in this chapter are not intended to be used to calculate loading but rather to build your intuition about how the design of a truss allows it to handle different loads.

CHAPTER 8   TRUSSES

MODELS USED IN THIS CHAPTER

This chapter uses two different OpenSCAD models. For more on 3D printing and creating and using math models in OpenSCAD, check out Chapter 1. Select 3D printable STL example files are included in the repository as well. The OpenSCAD models are

- `2DTruss.scad`: This model prints out an extrusion of a 2D truss. It is included as Listing 8-1.
- `tensegrityBeam.scad`: This model prints out beams to create tensegrity structures. It is included as Listing 8-2. You will need to print 9 copies of it (9 beams) to do all the projects in this chapter.

You will also need nine identical rubber bands to build the two described tensegrity structures. We suggest #14 or #16 rubber bands.

Engineering Background

You might first think of bridges when you hear the word *truss*. However, some bridges, like the Henley Street Bridge in Knoxville, Tennessee (see the foreground of Figure 8-1), have been designed with structural members that are arches or boxes. The bridge in the distance (the Gay Street Bridge completed in 1898), however, has a mostly triangular, 2D truss structure.

Trusses are usually defined as structures made up of thin members that are connected at joints in such a way that force is only applied to any member directly down its axis. That is, any given member is only in compression or tension and is not resisting any rotation around the joint. An ideal truss is, in fact, built with members connected by pin joints that are free to rotate within a plane.

CHAPTER 8 TRUSSES

Figure 8-1. *Two bridges in Knoxville, Tennessee*

In practice, trusses may not have their joints actually free to move. However, the overall structure may be designed so that very little rotational force should ever be applied to the joints, so that in practice it amounts to the same thing. In cases like this, joints are covered with a gusset plate or other reinforcement rather than a pin joint that can freely rotate.

Why Triangular Structures?

Trusses have been around in some form since people first made primitive bridges and houses. A triangle is the simplest structure that does not deform easily when a force is applied to one side. Consider the red box on the top of Figure 8-2 and imagine that its four joints are held together with pins that allow the members to rotate around the pin. If you apply a force to the top of it, it will eventually squash to one side or the other.

279

However, if there is a triangular cross-brace across the rectangular structure (as shown in the blue diagram), the structure is held rigid and cannot collapse without changing the length of at least one of the members. In order for the blue rectangle to collapse like the red one, the cross-brace would have to get longer. This cross-member is then experiencing tensile force. Collapsing in the opposite direction would instead apply a compressive force, because it would have to get shorter to allow a collapse. This is true even if the members are free to rotate around the joints holding them together.

Simplified analysis of trusses like these assumes the structure is two-dimensional, that its weight is negligible compared to the load it is supporting, and that all forces are applied along the axis of thin members in the 2D plane of the truss. If that is the case, you can figure out the force on each member by adding up the forces in two dimensions in each joint. This type of analysis also must allow for the fact that some joints will be experiencing some sort of external force (at the ends of a bridge, for example).

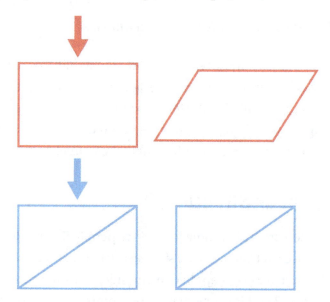

Figure 8-2. Forces in rectangular and triangular cross-section structures

Forces on Planar ("2D") Truss Members

When you add up the loads on each member, you wind up with several equations to balance the forces at each of the joints. Sometimes you can solve these sets of simple equations and get a useful result. Other times, you may have to make some assumptions or use more sophisticated techniques. Analysis allowing for the reality of 3D forces, materials properties, and so on is usually done with specialized software.

Under static loads, some members of a truss will be in tension (being pulled on each end) and others in compression (being pushed in along its length, as you would if you held a bar between your palms and pushed your palms inward). Some materials are stronger in tension or in compression, so if you know ahead of time what type of force that particular bar will have to take, you can design more efficiently.

Note We do not get into how to calculate these loads here since the subject is complex and more appropriate for follow-on projects. A first look at the issues for the 2D problem is described well in Wikipedia's "Truss" article, and you can find more resources by searching on "free body diagram truss bridge." R.C. Hibbeler's *Structural Analysis, 4th Edition* (Prentice Hall, 1999) is a good undergraduate structures textbook if you want to step up to the college-level version of this analysis.

The Space (3D) Truss

The 2D model of a planar truss applies if the loading applied to the truss is all in one plane. But suppose you want to hold up a more complex 3D load? A *space* (or 3D) truss will be needed, sometimes called a *space frame*.

CHAPTER 8   TRUSSES

These structures are often used on the back of billboards, as the underside of shade canopies, and in other places where a large, often flat object needs to be held up somehow without using heavy structures or too many supports in the middle of the span. A frame like this has hundreds or thousands of joints, so it becomes impractical to analyze them by writing equations for every joint. More complex methods and software are usually needed to analyze a space frame.

There are a lot of ways to make a simple space-filling truss with various construction toys, or marshmallows and uncooked spaghetti, or any number of other everyday objects. We do not think 3D printing will add a lot to that experience, so instead, we are going to show you how to create a somewhat more exotic space-filling structure using *tensegrity*.

Tensegrity Structures

Tensegrity structures use stiff members connected by strings or light cables; see the Wikipedia article "Tensegrity" for more. They are designed so that the cables are always in tension, while the stiff members are always in compression. Buckminster Fuller and artist Ken Snelson popularized the concept. Fuller patented a lot of key aspects of their architectural use in the 1960s through 1980s.

If you are interested in these structures and related ones like those in Fuller's work, you may find some good material for some interesting projects by building on our simple start here. Biologists find tensegrity structures interesting too, because bones and tendons are this type of structure.

The Models

In this chapter, we give you a few models to play with to build some intuition. There are a lot of other good ways to play with trusses (see "Where to Learn More"). Let's create a few unique models that take advantage of 3D printing's strengths.

CHAPTER 8 TRUSSES

2D Truss Model

Creating a 2D truss model is a little subtler than it seems. For a 2D truss model to be reasonably accurate, the joints between members need to be *pin joints*—that is, they need to rotate freely and not move into the third dimension when forces are applied. However, pin joints are difficult to do well with a consumer 3D printer (due to overhang issues and the relative weakness of interlayer bonds), particularly in a general model that allows the user to create relatively arbitrary trusses.

We decided that a student could functionally model a truss in any number of ways. However, most of these "toy trusses" are fragile and hard to play with too assertively. The model in Listing 8-1 and Figure 8-3 may look a little strange at first glance. We have added little feet at the ends so that you can stand it up on a table (or another flat surface) and add a load, as we do in Figure 8-4. Members compress and expand under a downward loading in the center, as we can see in the two overlaid images in Figure 8-5.

Note Since the "spring" members introduce some effects of their own, the joints are not perfect pin joints. However, we feel it gives some interesting qualitative insights. The compression and expansion under load are a little hard to see in Figure 8-3 but is more apparent when you load and unload the truss dynamically.

Listing 8-1. The 2D Truss (file 2DTruss.scad)

```
//OpenSCAD model of a truss with spring members
//File 2dTruss.scad
//(c) 2016-2024 Rich Cameron
//for the book 3D Printed Science projects, Volume 1
//Licensed under a Creative Commons, Attribution,
```

CHAPTER 8   TRUSSES

```
//CC-BY 4.0 international license, per
//https://creativecommons.org/licenses/by/4.0/
//Attribute to Rich Cameron, at
//repository github.com/whosawhatsis/3DP-Science-Projects

//thickness of the truss members, mm
beam_width = 3;
//Max length of the truss, mm
beam_length = 50;
//Width of the spring
spring_width = 15;
//Thickness of the spring members in x/y, mm
spring_thick = 1;
//Gap between turns of the spring
spring_gap = 3.3;
//number of turns in the springy part of each member
spring_turns = 9;
//depth of the truss in the "third dimension"
plane_thick = 20;
//height of the feet on the ends of the truss, mm
foot = 10;
//number of triangles in the truss (should be an odd number)
triangles = 5; //[1:2:15]

$fs = .2;
$fa = 2;
spring_pos = (spring_width + spring_thick + spring_gap)
 * 2 / 3;

//Extrude the 2D truss by plane_thick mm
linear_extrude(plane_thick) {
  for(i = [0:triangles - 1])
    translate(beam_length * [i / 2, (i % 2) * sqrt(3) / 2, 0])
```

CHAPTER 8　TRUSSES

```
      beam();
  for(i = [0:triangles])
    translate(beam_length * [i / 2, (i % 2) * sqrt(3) / 2, 0])
      rotate(60 * ((i % 2) ? -1 : 1))
        beam();
  for(i = [0, triangles + 1]) hull() for(j = [0, 1])
    translate([i * beam_length / 2, -foot * j, 0])
      circle(beam_width * 1.5);
}
//Create a truss beam using a spring
module beam(l = beam_length, w = beam_width) {
  spring((
    l - (spring_gap + spring_thick) * (spring_turns + 1)) / 2
  ) offset(w / 2) square([l, .01]);
  for(i = [0, 1]) translate([i * beam_length, 0, 0])
    circle(beam_width * 1.5);
}
//Create a spring for the truss beams
module spring(
  p = 0,
  w = spring_width,
  t = spring_thick,
  g = spring_gap,
  s = spring_turns
) {
  difference() {
    union() {
      if($children) for(i = [0: $children - 1]) children(i);
      for(i = [0:s]) translate([(i + .5) * (g + t) + p, 0, 0])
        mirror([0, i % 2, 0])
```

CHAPTER 8   TRUSSES

```
            difference() {
              translate([-.005, 0, 0]) offset(g / 2 + t)
                square([
                  .01,
                  (w / 2 - g / 2 - t) *
                    sin((i + .5) / (s + 1) * 180)
                ]);
              translate([0, -g - t, 0])
                square((g + t) * 2, center = true);
            }
         }
      for(i = [0:s]) mirror([0, i % 2, 0])
        translate([(i + .5) * (g + t) + p, 0, 0]) {
          translate([0, -w, 0]) offset(g / 2)
            square([
              .01,
              ((w / 2 - g / 2 - t) *
                sin((i + .5) / (s + 1) * 180) + w) * 2
            ], center = true);
          difference() {
            translate([0, -g - t, 0])
              square((g + t) * 2, center = true);
            for(j = [1, -1]) mirror([0, 1, 0])
              translate([j * (g + t), -w, 0]) offset(g / 2 + t)
                square([
                  .01,
                  w - g - t * 2 + w * 2
                ], center = true);
          }
        }
    }
  }
```

286

CHAPTER 8   TRUSSES

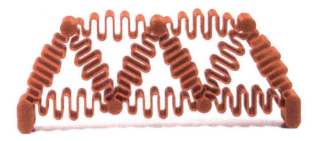

Figure 8-3. *The springy member 2D truss simulator*

Caution   Remember that this is a 2D truss, so resist the temptation to twist the model out of the plane. The "springs" are stiff in that direction to try and make it as close to 2D behavior as possible.

We encourage you to create a couple or three of these 2D trusses and play around with different support and loading scenarios. These are designed to be printed flat (with the layers parallel to the truss's plane), which will keep any layer-boundary effects or delaminations from affecting the 2D behavior. These might be good "pass-around" handouts to have in a class that is preparing to build trusses with construction toys or other materials.

Figure 8-4 shows the same model as in Figure 8-3, but now resting on its two feet and with one pressure point. Although there are a lot of distortions (notably that the endpoints on the bottom were not held in place), it is still clear to see how the triangular structure prevents the area right under that load from collapsing. You can also see how some of the members are in tension while others compress. In Figure 8-5, we have overlaid the compressed and original configurations of the truss to show the changes in the elements.

287

CHAPTER 8 TRUSSES

Figure 8-4. *Loading from above*

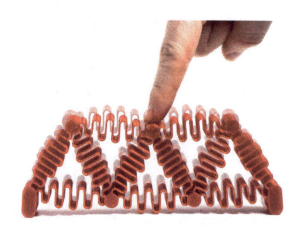

Figure 8-5. *Overlay of compressed and not*

The model in Listing 8-1 has a variety of parameters that allow you to vary how much of each member is made springy and the relative sizes of the members and features (as well as how many triangles you can have). The parameters are described in comments in the model. You may want to vary these and see what combination gets you closest to simulating the idealized truss with its pin joints and stiff members.

Printing the 2D Truss

The 2D truss should print fine without any special preparation. If, however, you do find the thin squiggly bits are not sticking enough, you can try printing with a *raft* (a layer of material on the printer platform—see Chapter 1). You should *not* use a brim, as it will affect the compressibility of the spring on one side if not removed perfectly. In that case, it will no longer act as a planar truss.

Tensegrity Structure Model

Next, we look at creating a demonstration of *tensegrity*—a way to use stiff members (like those in the trusses we just talked about) in combination with flexible members like cables, string, or rubber bands. The model in Listing 8-2 creates a single element. You should print three of them for the simple prism structure and six to make an icosahedron (20-sided polyhedron). You also will need three *identical* rubber bands for the prism and six for the icosahedron; #14 or #16 should work.

In general, you can find an optimum length for the cross-bracing strings (here, rubber bands) either by experimenting until you get things tight or by diving into the mathematics of these structures (www.tensegriteit.nl/e-simple.html). The rubber bands are the most forgiving way we found to build the structure.

In the section "Learning Like a Maker," we discuss how much iteration was needed to come up with this. In the next section, we talk you through assembling the 3-rod tensegrity prism (Figure 8-5) and, in the section after that, the general way you would extend it for the octahedron (Figure 8-6).

CHAPTER 8　TRUSSES

Figure 8-6. *The 3-rod tensegrity prism*

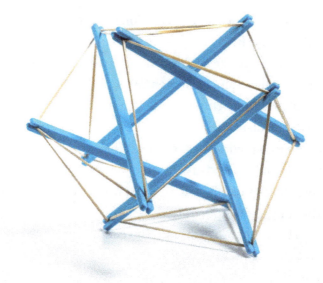

Figure 8-7. *The 6-rod tensegrity icosahedron*

Listing 8-2. Tensegrity Truss (file tensegrityBeam.scad)

```
//OpenSCAD model of a beam for building tensegrity structures
//File tensegrityBeam.scad
//(c) 2016-2024 Rich Cameron
```

CHAPTER 8 TRUSSES

```
//for the book 3D Printed Science projects, Volume 1
//Licensed under a Creative Commons, Attribution,
//CC-BY 4.0 international license, per
//https://creativecommons.org/licenses/by/4.0/
//Attribute to Rich Cameron, at
//repository github.com/whosawhatsis/3DP-Science-Projects

l = 120; // length of the rod
w = 6; // width of the rod, mm
h = 3; // depth in third dimension
hole = .5; // parameter to round off the edges of the slot

$fs = .2;
$fa = 2;

difference() {
  linear_extrude(h, convexity = 5)
    offset((w - hole) * .24) offset((w - hole) * -.24)
      difference() {
        square([w, l], center = true);
        for(i = [1, -1]) translate([0, i * l / 2, 0])
          square([hole, w * 2], center = true);
      }
  for(i = [1, -1], j = [0, 1]) {
    translate([0, 0, h / 2]) mirror([0, 0, j])
      translate([0, i * l / 2, -h / 2 - 1]) for(k = [0:.1:1]) {
        hull() for(k = [k, k + .1]) linear_extrude(2 - k)
          offset(1 - sqrt(1 - pow(k, 2)))
            square([hole, w * 2], center = true);
        hull() for(k = [k, k + .1]) linear_extrude(2 - k)
          offset(1 - sqrt(1 - pow(k, 2)))
            translate([0, -i * w, 0]) circle(hole);
        linear_extrude(h) {
```

291

```
            square([hole, w * 2], center = true);
            translate([0, -i * w, 0]) circle(hole);
        }
    }
  }
}
```

Printing the Tensegrity Elements

The tensegrity elements are very simple to print. The difficulties in these models lie in the assembly. The key there is to go about it slowly and carefully and to be sure all your rubber bands are as identical as possible. As a side note, beams may warp a bit if you store the tensegrity structure assembled for a long time (the rubber bands will also eventually break if stored under tension), so you may want to assemble it again if you want to display it.

Assembling the 3-Rod Tensegrity Prism

The tensegrity structures are a little tricky to assemble. We take you step by step through assembling a 3-member tensegrity prism. Print three of the members generated by the model in Listing 8-2 and label them, as shown in Figure 8-7.

Tip We labeled the rods in our test prism with numbers—odd number on top and even number at the bottom. Thus, we have rod 1–2, rod 3–4, and rod 5–6. You can see them in Figure 8-7. We refer to the rods by these numbers in what follows. Rod 1–2 is just *one* rod, not two—with end 1 and end 2.

CHAPTER 8   TRUSSES

Figure 8-8. *The labeled tensegrity rods*

Now put a rubber band over each of the rods, through the slit and over front and back (Figure 8-9). Check that the tension front and back is the same and that the rubber bands on all three rods are exerting about the same tension as each other (Figure 8-10). Otherwise, the structure may be hard to put together. You may need to test out a few rubber bands to get three that are the same as each other.

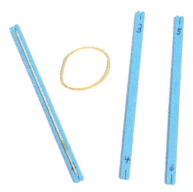

Figure 8-9. *A rod with a rubber band (and the unstretched band, for scale)*

CHAPTER 8   TRUSSES

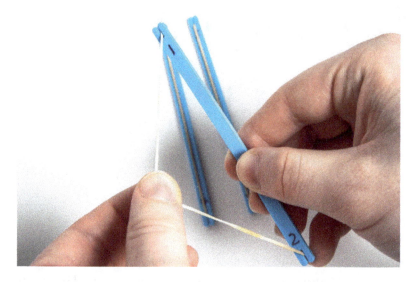

Figure 8-10. Testing the tension on the rubber band before starting to assemble

---

Tip   Keep the odd-numbered ends up and pointing away from you. Pull on rubber bands in the middle of their open space, as in Figure 8-9.

To start assembling, pick up rods 1–2 and 3–4. Pull up the band on 1–2 and hook the middle of rod 1–2's "front" band onto end 3 of rod 3–4 (Figure 8-11).

CHAPTER 8   TRUSSES

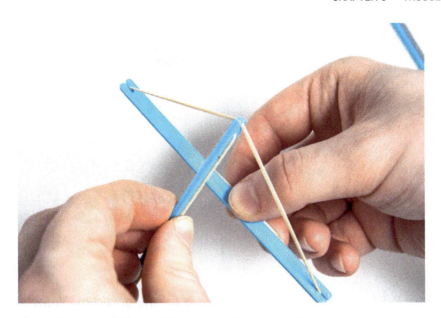

Figure 8-11. *Hooking the first two pieces together*

Next cross rod 3-4 over rod 1-2 so that the remaining rubber bands are on the "outside" of the crossed plastic bars (Figure 8-12).

295

CHAPTER 8   TRUSSES

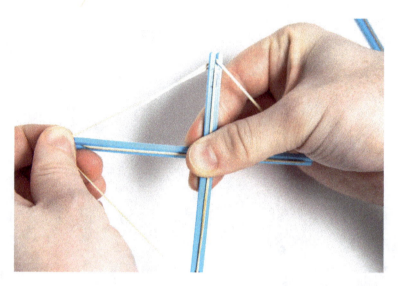

Figure 8-12. *Lining up rods 1-2 and 3-4 for the next step. Note that rod 3-4 has a rubber band still stretched across its "front" and rod 1-2 has one across its "back."*

Pick up rod 5-6 and put in the "V" of the space between 1-2 and 3-4. Use the band on one side of rod 5-6 to hook onto end 1 of rod 1-2 and end 4 of rod 3-4 (Figure 8-13).

CHAPTER 8 TRUSSES

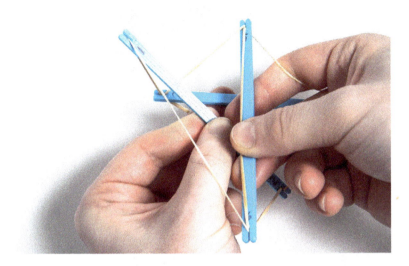

Figure 8-13. *Adding rod 5–6*

Now take the remaining rubber band on rod 3–4 and hook it on end 5, and take the remaining rubber band on rod 1–2 and hook it on end 6. You are done! Now you may need to fiddle with it a bit by gently removing the rubber bands from the end slits and adjusting where along the length of the rubber band they connect to square things up a bit.

The "top" should be a relatively flat equilateral triangle, as should the bottom. These should have single rubber band strands. The three vertical pieces should all be doubled up (Figure 8-14). You are looking "down" and through the top triangle here, which joins ends 1, 3, and 5. The bottom triangle joins ends 2, 4, and 6. The verticals run between ends 2 and 3, 4 and 5, and 1 and 6.

Figure 8-14. *The finished product, looking through the top triangle*

Hints for Assembling an Icosahedron

The icosahedron is a lot trickier. If you decide to assemble one, you may want to ask a friend (or two) to provide extra hands to hold the pieces in place. Figure 8-15 is another view of the icosahedron.

We do not walk through it step by step here, but we do give you the following hints so you can try it on your own:

- Every link is a single rubber band strand (as opposed to some of them being doubled up, as is the case with the prism).

- While in both cases there are two rubber bands going through each node, rubber bands go off in four directions from each node in the icosahedron but only in three directions (two of them doubled up together) in the prism.

CHAPTER 8 TRUSSES

Figure 8-15. *Another view of the icosahedron (compare to Figure 8-7)*

Note These trusses are not all that strong as printed, so be cautious if you want to see if you can put something light on top and see if it will stand. Try to avoid very old or previously used rubber bands (which might snap). Be careful not to shoot rubber bands off the ends of a beam during assembly.

THINKING ABOUT THE MODELS: LEARNING LIKE A MAKER

As with many of our models, we were surprised by how difficult it was to make even a simple 2D truss that would behave accurately. When we wrote the first edition of this book, we went through many iterations of how and where to place the springy part to mimic a pin joint as much as possible. We decided early on we did not want an actual pin joint since they are difficult to print well and hard to keep aligned so that all the forces stay in a 2D plane. The trusses

CHAPTER 8 TRUSSES

in Figure 8-16, except for the current (red) one at the bottom, were developed with past iterations of the model, not the one in Listing 8-1.

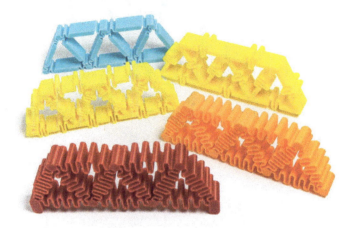

***Figure 8-16.** 2D truss iterations—moving around the "spring" area, changing its amplitude and how much of each member was "springy"*

For the tensegrity models, we had an assortment of failures. Our first model had a hole at the end of each member, rather than a slit at each end. This made it very hard to attach the string. After a lot of fussing with various types of connectors and clips to hold on the bands, Rich hit on the version you see earlier in the chapter.

The difficulty we encountered with assembling these structures is actually what convinced us that they were worth including. Removing any member makes a tensegrity become so unstable that it will completely fall apart, which makes it difficult to assemble because it will not be stable until it is finished.

Trying to use string was very frustrating, and we considered using wire for the tensile members so that they would hold their shape but decided that this was not true tensegrity. In the end, the way we stretched the rubber bands across the beams allowed the forces to be somewhat balanced, and while the assembly is still a bit tricky, this made it much more manageable.

CHAPTER 8   TRUSSES

Where to Learn More

There are various classic kids' engineering contests to build variants of bridges, towers, and trusses. If you have never seen a spaghetti and marshmallow (or toothpick and gumdrop) structure contest, search for it online and see how spectacular the results can be.

There are also a variety of bridge design games online. The group Engineering Encounters runs an annual bridge design contest complete with simulation software, at https://bridgecontest.org. There are also some fun, inexpensive games to allow students to play with creating reasonably realistic 2D trusses to hold up a roadway. Rich is partial to Poly Bridge (https://store.steampowered.com/app/367450/), and the same store has quite a few other options we have not tried, like the ones resulting from this search: https://store.steampowered.com/search/?snr=1_7_7_151_12&term=bridge+construction.

It is interesting (but scary) to read about bridges and other edifices that have failed, but it is a good way of learning about what might go wrong. To read about the great failures of bridges, you might read Henry Petroski's books, particularly *To Engineer Is Human* (Vintage Books, 1992) and the other books mentioned early in the chapter.

If you want to study a fun example of a moving machine with a lot of space-filling trusses, take a look at Theo Jansen's Strandbeests (www.strandbeest.com), which he has been "evolving" for years to move efficiently. His leg (truss) design is detailed in the Wikipedia article, "Jansen's linkage."

The tensegrity trusses we started building in this chapter (besides being featured in Buckminster Fuller designs) have appeared in sculptures and in some bridges, notably the Kirulpa Bridge in Brisbane, Australia. Try searching on "tensegrity architecture." Beyond the man-made, tensegrity structures are of interest in biology, particularly in the mechanics of how bones and tendons work together to create strong structures.

301

CHAPTER 8   TRUSSES

Teacher Tips

Broadly, looking at truss problems can be thought of as fitting into the general middle school or high school "Engineering Design" standards, found at www.nextgenscience.org/topic-arrangement/msengineering-design and www.nextgenscience.org/topic-arrangement/hsengineering-design, respectively.

Alternatively, this material might be used to facilitate discussions of forces and interactions in middle school, although these standards seem to be more about kinematics than the statics issues that we have explored in this chapter: www.nextgenscience.org/topic-arrangement/msforces-and-interactions.

Science Fair Project Ideas

For a relatively simple project, play with some of the parameters for the 2D truss structure. You might see how varying the parameters changes behavior to more closely mimic an ideal pin joint. When you are looking at behavior, we recommend against trying to stress 3D printed models to failure. The plastic is fairly brittle, and the layer lines may put some stress in places you did not expect. Wear eye protection if you think things may snap.

More complex tensegrity projects are good options for science fair projects, too. How much load can these structures hold? What types of forces are they good at withstanding, and what types of forces are likely to make them collapse? Perhaps you can look into a biological tensegrity structure and figure out how to build something that will mimic the real thing well enough to gain some intuition.

Summary

This chapter developed models of trusses, structures that carry loads by strategically using triangular arrangements of members. The chapter showed you how to print a 2D truss and the elements of a tensegrity truss. We also looked at applications of trusses and ideas for more explorations.

CHAPTER 9

Gears

In this final chapter (new for the second edition of this book), we expand on the simple machines we learned about in Chapter 5. We focused there on getting mechanical advantage by exerting forces using levers, pulleys, inclined planes, wheels and axles, and wedges. In Chapter 5, we thought mostly about linear motion: exerting forces to make something move in a straight line.

Wheels and axles are rotating but are usually rolling under something that is going in a straight line. Levers, however, rotate about a fulcrum to move something. We did not put it that way back in Chapter 5, but levers are creating mechanical advantage in a way that produces a force at the end of the lever arm, known as *torque*.

More generally, torque is a force applied by a rotating mechanism at some distance out from the axis of rotation. Torque might be applied by a rotating shaft (like a spinning motor shaft) or at the outer radius of something that is turning, like one gear turning another (Figure 9-1), or as we mentioned, as a lever pivots around a fulcrum.

Torque has the units of force times a distance (newton-meters in SI units) since it is a force multiplied by its distance to the center of rotation. Strictly speaking, when you learn calculus, you will find out that torque is defined as a special kind of multiplication called a "cross product" which considers the direction of rotation and the force. For now, we will think of it as ordinary multiplication to get some intuition. The farther from the

CHAPTER 9 GEARS

axis of rotation the force is applied, the bigger the torque (just like longer levers). This means that, as depicted in Figure 9-1, a small force applied at the end of a lever can produce the same torque as a larger force applied closer to its fulcrum.

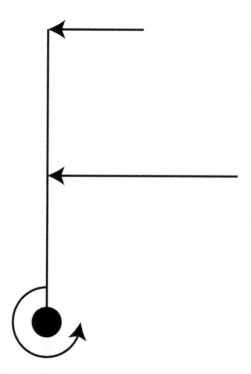

Figure 9-1. Applying torque

Various mechanisms can create torque or convert rotational to linear motion or vice versa. They also might change the mechanical advantage of a device by applying a smaller force for a longer distance. There are many different types, but we will focus on gears. Let's see some of the types of gears and how and why we might use them. We generate a simple gear model here, but we encourage you to look around and see how many useful devices are variations on these.

CHAPTER 9 GEARS

MODELS USED IN THIS CHAPTER

This chapter uses one OpenSCAD model described here and points to one in another collection. For more on creating and using math models in OpenSCAD and 3D printing in general, refer to Chapter 1. Select 3D printable STL example files are included in the repository as well. The models for this chapter are

- `gears.scad`: Creates an assembly of four gears
- A planetary gear bearing at www.youmagine.com/designs/quick-print-gear-bearing

Gears

If we wanted to have a shaft turning one wheel to rub against another so that it, too, would rotate, smooth wheels might slip instead of rolling against each other. Thousands of years ago, people realized that if they put interlocking protrusions (referred to as *teeth* or *cogs*) around the edges of both wheels, they could rotate together without slipping. These wheels with teeth are *gears*, pairs of wheels with teeth that mesh together. Check out the Wikipedia entry "Gears" for a very thorough overview of the many variations.

If one gear is larger than the other, the smaller wheel will turn more than once for each rotation of the larger one. This means that gears are a way of turning slow rotation into faster or vice versa. If you have a larger and smaller gear turning together, the teeth on both gears need to mesh together smoothly. This means the teeth on both gears need to have the same spacing. This spacing is known as the gear's pitch, just like the screw pitch we learned about in Chapter 5 but measured along the gear's circumference. You could think of a gear tooth as a sort of wedge (see Chapter 5), exerting transient forces to minimize sideways slip. Finally, one gear of a pair will turn clockwise and the other counterclockwise (roll two coins against each other to see why).

We need to introduce a few definitions to be able to talk about how gear meshing happens. First, what is the "radius" of a circular gear? Options might be the radius of the wheel including the maximum height of a tooth or alternatively the radius without the teeth. As it turns out, there is a convention of using a pitch circle or reference circle. Imagine you have two gears meshing and you drew a circle around the center of each gear, such that the two circles just touched at one point. Those circles would be the pitch circles, which will typically be the radius without the teeth plus about half the height of a tooth (Figure 9-2). When we talk about the radius of a gear going forward in this chapter, we mean the pitch radius.

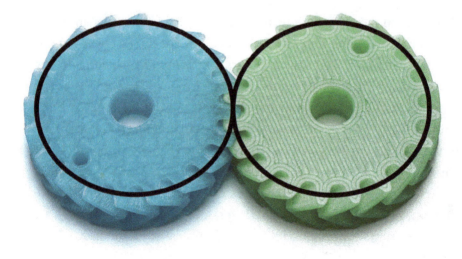

Figure 9-2. *A pair of gears showing pitch radius*

Gear Ratio

For gears measured in the metric system, the size of the teeth is specified by the gear's module. The module of a gear is the diameter of the reference circle divided by the number of teeth. For gears measured in inches, otherwise known as imperial gears, it is the inverse and called diametral pitch. We will use the metric convention in this chapter. We can think of the metric version as the amount of space allocated to one tooth plus any gap between teeth. The imperial version is the number of teeth per inch.

To mesh, two gears need to have the same module as each other, or the teeth would be different sizes. Each gear needs to have a whole number of teeth, assuming it is a gear designed to rotate a full turn. That means that once you pick a number of teeth on one gear, the diameter, and the size of the teeth, there are constraints on the exact size of the second gear and on the distance between the gears' centers. To have even wear on all teeth, it is also usually recommended that the number of teeth on the two gears each be a (different) prime number or at least two numbers that do not share a prime factor.

A gear being turned by another gear with half its radius will turn half as fast but deliver twice the torque, the force delivered by a rotating mechanism. Torque is the product of the radius of rotation times the applied force, so a larger radius gives more torque. Levers (Chapter 5) produce torque as well, which is why a longer lever produces more torque.

If a motor is used to turn a gear, the resulting torque exerts a force at the gear's teeth proportional to the gear's radius. The gear contacting it experiences the same force. However, its radius is different. Since the two gears cannot slip, the speed along the reference circles of the two must be the same. However, since the circumference of the second motor is different, the torque will be the ratio of the radii of the two gears. The bigger gear of the pair will turn more slowly but exert more force.
The ratio of the radii is called the gear ratio. They are often expressed in a format like 1:3.

CHAPTER 9   GEARS

It would be inconvenient if everyone needed to design a custom motor to spin at the particular speed a design required, and most electric motor designs perform most efficiently at very high rotational speeds. Fortunately, gears let you get around that, within the limits of meeting all the constraints mentioned earlier. Clocks were an early application of precision gears, and a lot of patience and skill were needed to make tiny gears until adequate metal cutting processes came along. Some watch parts are still handcrafted for watches that have mechanical parts (unlike digital watches, which typically contain no moving parts, except maybe for buttons). An hour, minute, and second hand all need to go around the dial at different speeds, which can be done with gearing. The gears are so tiny that they have historically been designed to turn on small jewels to keep the friction low.

The Exploratorium museum in San Francisco has an exhibit by artist Arthur Ganson that is a long gear train. The gear at one end spins visibly, driven by a motor. The one at the far end, with many intermediary gears, has been slowed down so much that it can be embedded in concrete since it will take billions of years to complete one full rotation! Check it out if you find yourself there, or see it on the artist's website, `www.arthurganson.com`. Look on the site under `Sculpture ▶ Machine with Concrete`, and check out some of his other fun mechanisms too.

Types of Gears

There are many different types of gears, with custom styles for specialized applications. We usually think of gears as round and more or less 2D. Ones that are in this category are the *spur gear* (teeth that are cut so that they are symmetrical as they rotate and the same thickness as the round part of the gear) and *helical gear* (teeth around the edge of the gear at an angle). The gears in Figure 9-2 are called *herringbone gears*, and we will meet them when we talk about the model you can make.

Gears also do not need to be round. Gears that are elliptical or shaped like a seashell (*nautilus gears*) can create an erratic motion that might be needed for some applications. Search on "nautilus gear sculptures" for some mesmerizing applications.

Another type of gear is a *rack*, which is essentially a linear gear. You can think of it as a gear that has been "unrolled" or as a segment of a gear with an infinite radius. Another gear will run along this gear, like a wheel running along a road (but with teeth, so the wheel will not slip). This arrangement is sometimes called a *rack and pinion*: the rack is the straight part, and the turning part is the pinion. A variation of this is the worm, which looks like a screw thread. As a worm spins against a gear, each full turn of the worm causes the gear to advance by one tooth. This can produce a high gear ratio in a small space, but the efficiency is low due to the friction of the worm threads sliding sideways along the gear teeth.

Note The term pinion is also used beyond its use in "rack and pinion." Pinions are always round gears with external teeth (i.e., not racks, internal gears, or worms), and the term normally refers to the smaller of a pair of gears. Pinions are often produced as long cylindrical rods (called *pinion stock*) with teeth around the circumference that run along the length. This contrasts with the larger "wheel" gear that it turns, which is more likely to be cut out from material that is only about as thick as the gear. In clocks and watches, the (sometimes paper-thin) wheel gear is probably cut (or possibly even punched) out of thin sheet metal. That is assuming the gears are not plastic, as mass-produced plastic gears are typically individually molded, regardless of their size.

CHAPTER 9 GEARS

There are also *internal gears*, where the teeth are on the inside of a circle, also called annular gears and in some cases ring gears. One or more smaller gears will run around the inside to fit some required motion. We will see a model of a *planetary gear* (where several small gears turn inside an internal gear) later in this chapter.

Sprockets

Sometimes, instead of running two gears directly against one another, a chain is used to span the distance between them. Instead of gears, this requires a superficially similar-looking device called a *sprocket*. Sprockets have teeth like gears but with a different shape that is optimized for meshing with links to the chain, rather than with one another. An obvious example is a bicycle, where the human rider moves a crank up and down by pedaling. Sprockets attached to the pedals and ones attached to the wheels are separated by a chain.

A bicycle's "lower gears" for going uphill use the bigger diameter sprockets on the back wheel, but the smaller ones on the gears are directly driven by the pedal. This seems counterintuitive, but we want each turn of the crank to turn the wheel a little less when we must exert a lot of force to move the wheel. If we are pedaling along on level ground, we want each push of the pedal to move us a larger distance since the force needed is less. On a bicycle, these sprockets are usually in a cluster (Figure 9-3) and with additional mechanisms to allow the rider to jump the chain from one sprocket to the next, changing the effective gearing ratio.

CHAPTER 9 GEARS

Figure 9-3. Bicycle sprocket cluster, back wheel

Sprockets and chain drives are closely related to timing belts, which work similarly, but replace the chain with a rubber belt with teeth molded into it. These mesh with matching timing pulleys. 3D printers typically use timing belts to run their faster-moving axes (while screw drives are more common for the slower z axis). By clamping onto one point on the belt, a spinning timing pulley produces linear motion along the belt's span (Figure 9-4).

CHAPTER 9   GEARS

Figure 9-4. Timing belt on a 3D printer

Gear Models

Gears have been a popular 3D printer application since the early days of consumer 3D printers when early 3D printer makers had to make their own extruder gears. There are many print-in-place sets of gears in online repositories, which, as the name implies, are designed to have just enough gaps between interlocking gears so that a whole assembly will print together with no assembly required. This allows sets of gears that might be impossible to assemble to be printed, too. In this chapter, we give you a very simple model to get some intuition about gearing and point you to a freely available print-in-place gear toy.

If you go to a repository site like Printables (`www.printables.com`) and search on "gears," you will get many hits including some entire collections. You might find the best fidget toy ever, too. We give further suggestions in the "Where to Learn More" section at the end of this chapter.

One advantage of printing your own gears is that you can create the gear ratio you need for a project. If you are working with a kit that includes some gears, you are stuck with whatever ratio their limited set supplies. If you know the module of your gears (as well as some other aspects of the gear's design), in principle, you can design gears that match extant ones. However, metal and molded plastic gears are often made with teeth too fine for a filament printer to match, so you might need to buy into making all or none of the gears for your project.

Gear Set Model

We have created a very simple gear model, gears.scad, that allows you to see how using a few gears in succession multiplies the gear ratio. The model has four pieces, as shown in Figure 9-5: a base, a pair of gears in a 2:1 ratio that are connected to run on the same shaft, and two freestanding gears, each of which is the same size as one of the connected gears. The gears are herringbone gears, called that because of their chevron-shaped teeth. They 3D print well and make it obvious in a photo which direction each gear is turning. The way herringbone gears interlock also prevents them from slipping along one another axially. This makes assembly a bit trickier but helps hold the whole mechanism together once assembled. You can prove to yourself by placing the small gear along the radius of the bigger one that it is indeed half the radius.

CHAPTER 9   GEARS

Figure 9-5. *Pieces of the gear model*

Assembling the model can be a bit tricky and is easiest to do upside down. Start by placing the smallest (blue) gear in the palm of your hand, with the crank side down (Figure 9-6). Hold the red double gear above it, with its smaller gear facing up, and mesh the teeth of the red larger gear with the blue one, making sure that the small hole through the red gear lines up with the hole through the crank (Figure 9-7).

CHAPTER 9 GEARS

***Figure* 9-6.** *One of the smaller gears, with a handle*

***Figure* 9-7.** *Meshing the blue gear with the larger red gear*

CHAPTER 9   GEARS

Adjust your fingers to hold these two pieces in this position, then add the third (green) gear. Its central hole will line up with the one on the small blue gear, and the smaller hole should align with the small holes on the other pieces. You may need to flip the gear over so that the herringbone teeth mesh properly. If you have trouble holding the gears in place at this step, you can temporarily insert a short piece of filament (about 30 mm, or a little over one inch long) through the small holes to lock the pieces together and help keep them aligned (Figure 9-8). Then just slide the base down on top of them, making sure that the shorter shaft goes through the double gear (Figure 9-9). Slide it all the way down, then turn it over. If you used a piece of filament to help with assembly, remove it to allow the gears to turn. Notice that in Figure 9-8, the herringbone teeth point clockwise around the green gear and counterclockwise on the red one.

Figure 9-8. *Meshing the green gear with the smaller red one*

CHAPTER 9   GEARS

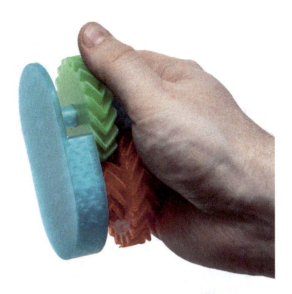

Figure 9-9. *Stacking the gears onto the base (side view)*

Figure 9-10. *The assembled gears*

Next, turn the small gear on top 180 degrees so it points the other way. The big gear turns a quarter turn, as we would expect from the 2:1 ratio of their diameters. If we keep going and turn the small gear one full revolution, the connected gears go a half revolution and the big

319

CHAPTER 9 GEARS

independent gear (which is coupled by another 2:1 ratio of the gears on the bottom) will have only gone one-quarter as far around, as we can see from the positions of the hole in each gear (Figure 9-11). Try out different amounts of rotation to see how this works. After four full rotations of the small gear, they will return to where they started, with the small holes aligned (back to the configuration in Figure 9-10). The model can be adapted by changing the parameters in Table 9-1, and you can see the listing of the model in Listing 9-1.

Figure 9-11. *Gears after one full rotation of the top gear*

CHAPTER 9   GEARS

Table 9-1. *Gear Model Parameters*

Variable	Default Value and Units	Meaning
shaft	10 mm	Diameter of the gear shafts
mod	2.5 mm	Gear module
teeth	[10, 20]	Number of teeth for the smaller and larger gears. This determines the gear ratio. These sizes (multiplied by the module) also determine the size of the base.
pressure_angle	14.5 degrees	Angle used for generating the shape of the gear teeth
height	mod * 5	Height of each gear (axial dimension)
base	3 mm	Base thickness
clearance	0.5 mm	Additional radius for holes around the shafts to allow them to spin freely

Listing 9-1. Listing of the File gears.scad

```
//OpenSCAD model to create a two-stage gear reduction
//File gears.scad
//(c) 2024 Rich Cameron
//for the book 3D Printed Science projects, Volume 1
//Licensed under a Creative Commons, Attribution,
//CC-BY 4.0 international license, per
//https://creativecommons.org/licenses/by/4.0/
//Attribute to Rich Cameron, at
```

CHAPTER 9 GEARS

```
//repository github.com/whosawhatsis/3DP-Science-Projects

//Diameter of the gear shafts
shaft = 10;
//Gear module
mod = 2.5;
//Number of teeth for the smaller and larger gears
teeth = [10, 20];
//Angle used for generating the shape of the gear teeth
pressure_angle = 14.5;
//Height of each gear (axial dimension)
height = mod * 5;
//Base thickness
base = 3;
//Additional radius for holes around shafts
clearance = .5;

$fs = .2;
$fa = 2;

distance = mod * (teeth[0] + teeth[1]) / 2;

module base() {
  linear_extrude(base) hull() for(i = [0, 1])
    translate([i * distance, 0, 0])
      circle(mod * max(teeth) / 2 + mod);
  linear_extrude(base + 1) for(i = [0, 1])
    translate([i * distance, 0, 0]) circle(shaft / 2 + 1);
  linear_extrude(base + 1 + height) offset(2) offset(-2)
    difference() {
      hull() for(i = [0, 1]) translate([i * distance, 0, 0])
        circle(mod * max(teeth) / 2 + mod);
      circle(mod * teeth[0] / 2 + mod + clearance * 1.5);
```

```
      translate([distance, 0, 0])
        circle(mod * teeth[1] / 2 + mod + clearance * 1.5);
    }
    linear_extrude(base + 2 + height * 2) circle(shaft / 2);
    linear_extrude(base + 3 + height * 3)
      translate([distance, 0, 0]) circle(shaft / 2);
}
base();
%translate([distance, 0, base + 1])
  rotate(90) gear3();
%translate([0, 0, base + 2 + height * 2])
  rotate([180, 0, 90]) gear2();
%translate([distance, 0, base + 3 + height * 3])
  rotate([0, 180, 90]) gear1();

translate([distance * 2 + mod * 2 + 1, -distance / 4, 0])
  gear1();
translate((mod * max(teeth) + mod * 2 + 1) * [1, 1, 0])
  gear2();
translate([0, mod * max(teeth) + mod * 2 + 1, 0])
  gear3();

module gear1() {
  linear_extrude(height) difference() {
    union() {
      circle(mod * teeth[0] / 2 + mod);
      hull() for(i = [0, 1]) translate([0, i * distance, 0])
        circle(mod * (teeth[1] - teeth[0]) / 8 + 1);
    }
    circle(shaft / 2 + clearance);
    translate([0, distance / 2, 0])
      circle(mod * (teeth[1] - teeth[0]) / 8 - 1);
```

CHAPTER 9 GEARS

```
    }
    linear_extrude(height * 2 + 2) difference() {
      circle(shaft / 2 + clearance + 1);
      circle(shaft / 2 + clearance);
    }
    intersection() {
      cylinder(r = teeth[0] * mod / 2 + mod, h = height * 2 + 1);
      translate([0, 0, height * 1.5 + 1])
        herringbone(height * 2, -360 / teeth[0] * 2)
          gear(teeth[0], mod, shaft / 2 + clearance);
    }
}

module gear2() {
  difference() {
    translate([0, 0, height / 2]) rotate(180 / teeth[1])
      herringbone(height, 360 / teeth[1])
        gear(teeth[1], mod, shaft / 2 + clearance);
    rotate(180) translate([0, -distance / 2, 0])
      linear_extrude(height)
        circle(mod * (teeth[1] - teeth[0]) / 8 - 1);
  }
  intersection() {
    linear_extrude(height * 2 + 1)
      circle(teeth[0] * mod / 2 + mod);
    translate([0, 0, height * 1.5 + 1]) rotate(180 / teeth[0])
      herringbone(height * 2, 360 / teeth[0] * 2)
        gear(teeth[0], mod, shaft / 2 + clearance);
  }
}
```

CHAPTER 9 GEARS

```
module gear3() difference() {
  translate([0, 0, height / 2])
    herringbone(height, 360 / teeth[1])
      gear(teeth[1], mod, shaft / 2 + clearance);
  translate([0, distance / 2, 0]) linear_extrude(height)
    circle(mod * (teeth[1] - teeth[0]) / 8 - 1);
}

module herringbone(thick = 12, angle = 0)
  for(m = [0, 1]) mirror([0, 0, m])
    linear_extrude(thick / 2, twist = angle, convexity = 5)
      children();

module gear(teeth, mod = 2, bore = 0) {
  baser = teeth * cos(pressure_angle);
  difference() {
    scale(mod / 2) offset(-1.2) offset(1.2) union() {
      circle(teeth - 2.5);
      for(tooth = [0:teeth]) rotate(tooth * 360 / teeth)
        intersection() {
          circle(teeth + 2);
          intersection_for(i = [0, 1]) mirror([i, 0, 0])
            rotate(-360 / teeth / 4)
              polygon([
                for(i = [-teeth / 10:.1:teeth / 2])
                  sqrt(sign(i) * (2 * PI * i)^2 + baser^2) * [
                    -sign(i) *
                    sin(
                      i * 360 / baser -
                      atan2(2 * PI * i, baser)
                    ),
                    cos(
```

```
                i * 360 / baser -
                atan2(2 * PI * abs(i), baser)
            )
        ]
    ]);
   }
  }
  circle(bore);
 }
}
```

Planetary Gear Model

A *bearing* is a mechanism using gears (or sometimes small balls, called ball bearings) that supports something that needs to turn freely. *Planetary gear bearings* have multiple gears around a central one, like planets around the Sun. (That is, if all the planets were in orbits the same distance from the Sun as each other.)

Rich created a planetary gear bearing (Figure 9-12) based on one by Emmett Lalish at www.thingiverse.com/thing:53451. The instructions for Rich's version are at www.youmagine.com/designs/quick-print-gear-bearing. These gears print in place very fast relative to other 3D printed gears and are fun for demonstrations. They are a little tricky though—you are printing the *hollow spaces inside* the model, not the model surface itself. This can be done with creative use of slicing program settings. The gears come off the printer ready to turn once you wiggle them a little to free them up. This print is a fun fidget spinner!

Figure 9-12. Rich's quick print gear (prints assembled, just as shown)

Where to Learn More

There are many places to learn more about gears and other mechanisms. There are many videos on YouTube, notably Nguyen Duc Thang's channel that has thousands of animated mechanisms (www.youtube.com/@thang010146). There are also many videos about how various industrial processes work on YouTube.

If you want to do more hands-on activities, competitive robotics teams need to learn a lot about gearing. LEGO Technic kits (meant for basic robot competition) are a good place to start, or there are a variety of more generic and less pricey robot kits. VEX and FIRST Robotics competitions are a higher end way to get into using the material in this chapter.

Teacher Tips

Like the material in Chapter 8, this material can be thought of as fitting into the general middle school or high school "Engineering Design" standards, found at www.nextgenscience.org/topic-arrangement/msengineering-design and www.nextgenscience.org/topic-arrangement/hsengineering-design, respectively. Alternatively, this material might be used to facilitate discussions of forces and interactions to extend a middle school discussion of simple machines, which might fall under www.nextgenscience.org/topic-arrangement/msforces-and-interactions.

If you have access to LEGO Technic materials (like LEGO Mindstorms EV3 or Spike sets) or VEX or FIRST robotics materials, you might follow some of the projects there and have students analyze gear ratios. Consider ways to go beyond the roving robot and find ways to use the motors and gears to give more intuition. There are also many projects out there to make things that move with Arduino microprocessors and hobby servos, which incorporate (or can be interfaced with) gears.

Science Fair Project Ideas

The ideas in this chapter would probably fall into physics or engineering categories at a science fair. Projects to play with gear trains and figure out interesting ways to measure force transmitted by various arrangements might be interesting (by lifting small weights, for example). Finding ways to measure the efficiency (although requiring some thought) of a progressively more complex machine would require computing expected torque and actual torque.

Summary

In this chapter, we explored how gears work to produce mechanical advantage and how they can be used to slow, speed up, or change the direction of rotational motion, with subsequent effects of the torque generated. We saw that there are many different classes of gears that can serve differing mechanical requirements. We explored simple 3D printed models to gain intuition about these ubiquitous devices.

CHAPTER 9   GEARS

A Few Last Words About Making Things

At the end of this final set of models, we would like to thank you for reading our book and (we hope) making things. There are many sophisticated computer programs out there to simulate and calculate answers to many different engineering problems.

But to learn to design something new—and to know when the programs might be outside of their assumptions—you need to develop intuition. We hope that these simple models have served that purpose for you. In some cases (most particularly these last ones), you may learn more by thinking about how these models do not fit reality and/or the common idealized models. But how could you make them better? Are the ways these little models work different from reality in a way that teaches you something else?

We hope that if nothing else, we have imparted a reflex to say, in the face of big unknowns, "I do not know what is going on here, but let's see if we can build a few simple models to find out!"

APPENDIX

Links

This appendix aggregates all the links in the book in one place for convenient reference. If a link appeared in more than one chapter, it is listed here under the chapter in which it first appeared.

About the Authors

Nonscriptum LLC:
    www.nonscriptum.com

Introduction

Creative Commons CC-BY license:
    http://creativecommons.org/licenses/by/4.0/

Chapter 1: Math Modeling with 3D Prints

OpenSCAD 3D modeling software:
    www.openscad.org
    OpenSCAD manual:
    https://en.wikibooks.org/wiki/OpenSCAD_User_Manual/Mathematical_Functions
    The authors' repository of models in this book:

APPENDIX LINKS

https://github.com/whosawhatsis/3DP-Science-Projects

Prusa Research software:

www.prusa3d.com

The author's *Make: Calculus* model repository:

https://github.com/whosawhatsis/Calculus

The Khan Academy online learning:

www.khanacademy.org

Elizabeth Denne's math modeling sites:

http://mathvis.academic.wlu.edu

Mathematical sculptor Bathsheba Grossman:

www.bathsheba.com

Mathematical sculptor Henry Segerman:

www.shapeways.com/shops/henryseg

Paul Nylander math models:

http://bugman123.com

Miscellaneous math models online:

www.thingiverse.com, www.printables.com

Chapter 2: Light and Other Waves

Very Large Array radio telescope:

www.vla.nrao.edu

Next Generation Science Standards:

www.nextgenscience.org

Next Generation Science Standards related to waves:

www.nextgenscience.org/msps-wer-waves-electromagnetic-radiation

www.nextgenscience.org/4w-waves

www.nextgenscience.org/hs-ps4-3-waves-and-their-applications-technologiesinformation-transfer

www.nextgenscience.org/4-ps4-1-waves-and-their-applications-technologies-information-transfer

APPENDIX   LINKS

Chapter 3: Gravity

Gravitational potential well, from cartoon strip XKCD:
   https://xkcd.com/681/
   NASA exoplanets site:
   https://science.nasa.gov/exoplanets/
   Next Generation Science Standards applying to gravity:
   www.nextgenscience.org/msess-ss-space-systems
   www.nextgenscience.org/ms-ess1-2-earths-place-universe
   Algol's brightness:
   www.skyandtelescope.com/astronomy-blogs/behold-algol-star-secret/

Chapter 4: Airfoils

The Incomplete Guide to Airfoil Usage:
   http://m-selig.ae.illinois.edu/ads/aircraft.html
   The Incomplete Guide to Airfoil Usage links page:
   http://m-selig.ae.illinois.edu/ads.html
   Site about airfoils in general:
   http://airfoiltools.com
   Aeronautics for kids:
   www.grc.nasa.gov/WWW/K-12/airplane/bgt.html
   Aeronautics for teachers:
   www.grc.nasa.gov/WWW/k-12/airplane/topics.htm
   DIY student wind tunnel design:
   https://sciencebuddies.org/science-fair-projects/references/how-to-build-a-wind-tunnel
   Simpler DIY student wind tunnel design:
   www.instructables.com/id/DIY-Wind-Tunnel-20-Project-Paperclip/

APPENDIX   LINKS

Science standards about energy and motion:
www.nextgenscience.org/msps2-motion-stability-forces-interactions
www.nextgenscience.org/msps3-energy

Chapter 5: Simple Machines

Forces and Interactions science standards:
www.nextgenscience.org/msps2-motion-stability-forces-interactions
Energy standards:
www.nextgenscience.org/msps3-energy

Chapter 6: Plants and Their Ecosystems

Golden ratio:
www.mathsisfun.com/numbers/golden-ratio.html
K-2 science standards about interrelationships between plants, animals, and environment:
www.nextgenscience.org/topic-arrangement/kinterdependent-relationships-ecosystems-animals-plants-and-their-environment
Middle school standards:
www.nextgenscience.org/pe/ms-ls2-5-ecosystems-interactions-energy-and-dynamics
High school:
www.nextgenscience.org/pe/hs-ls2-7-ecosystems-interactions-energy-and-dynamics
Biology citizen science site:
www.inaturalist.org

Chapter 7: Molecules

Royal Chemical Society interactive periodic table:
www.rsc.org/periodic-table
PubChem's table:
https://pubchem.ncbi.nlm.nih.gov/periodic-table/
Site to look for scientific articles:
https://scholar.google.com
ChemWiki:
http://chemwiki.ucdavis.edu/Theoretical_Chemistry/Chemical_Bonding/Valence_Bond_Theory
The International Union of Pure and Applied Chemistry (IUPAC) periodic table:
https://iupac.org/what-we-do/periodic-table-of-elements/
Molecular geometries resulting from different types of bonds:
www.kidzsearch.com/wiki/Orbital_hybridization
This video explains hybridization in terms of energy:
www.youtube.com/watch?v=HKyobMewXBw
These videos use balloons to model bonding:
www.youtube.com/watch?v=bOKvfvJi-vk
www.youtube.com/watch?v=KbOmxAMHnfE
Chem LibreTexts:
https://chem.libretexts.org
Science standards, chemical reactions:
www.nextgenscience.org/topic-arrangement/hschemical-reactions
Matter and Its Interactions, science standards:
www.nextgenscience.org/dci-arrangement/ms-ps1-matter-and-its-interactions

APPENDIX   LINKS

Chapter 8: Trusses

Tensegrity in-depth info site:
   www.tensegriteit.nl/e-simple.html
   Bridge contest:
   https://bridgecontest.org
   Poly Bridge game site:
   http://store.steampowered.com/app/367450/
   Site to purchase other games:
   http://store.steampowered.com/search/?snr=1_7_7_151_12&term=bridge+construction
   Strandbeest main page:
   www.strandbeest.com
   Engineering Design science standards:
   www.nextgenscience.org/topic-arrangement/msengineering-design
   www.nextgenscience.org/topic-arrangement/hsengineering-design
   Standards for forces and interactions in middle school:
   www.nextgenscience.org/topic-arrangement/msforces-and-interactions

Chapter 9: Gears

Website with fun machines:
   www.arthurganson.com
   Rich's Quick Print Gear Bearing:
   www.youmagine.com/designs/quick-print-gear-bearing
   Emmett Lalish gear:
   www.thingiverse.com/thing:53451
   Nguyen Duc Thang's channel that has thousands of animated mechanisms:
   www.youtube.com/@thang010146

Index

Numbers

3D math functions, 2–13
3D printing, 23–34
    archives and repositories, 33–34
    considerations, 33
    filament-based 3D
        printing, 23–33
    file types, 11
    mathematical functions, 1, 5
    OpenSCAD
        download, 3–4
        editing, 4
        ideosyncracies, 13
    raft, 29
    support, 25–28

A, B

Airfoils, 333
    camber line, 107
    chord, 107
    flight forces
        drag, 104
        gravity, 104
        lift, 104
        thrust, 104

NACA airfoils
    angle of attack, 111
    camber line, 117
    dihedral features, 112
    four-digit profiles, 114
    swept wings, 109
    tapered wings, 109
    thickness equation, 118
    in World War II
        airplanes, 112
printing tips, 143
thickness, 107
3D printed airfoil models
    measuring lift, 138
    sting, 127
    wind tunnel, 113
Algol model, 79

C

Carbon atom model, 253
Chain drives, 313
Conservation of energy
    equation, 84
Coordinate system and
    conventions, 60
Covalent bond, 241

INDEX

Crystal structures, molecules
  ice 1c, 267
    assembling, 269
  ice 1h, 264
    assembling, 266
  ice-nine, 269

D

Diametral pitch, 309

E

Earth-Moon system model, 71
Electron clouds, 242
Envelope model, 60

F

Fourier Transforms, 61
Fraunhofer Diffraction, 53
Function, 5

G

Ganson, Arthur, 310
Gear
  annular, 312
  in clocks, 310, 311
  helical, 310
  herringbone, 310, 315
  imperial, 309
  internal, 312
  module, 309, 321
  nautilus, 310
  pitch, 307
  pitch circle, 308
  planetary, 307, 312, 326
  reference circle of, 308, 309
  ring, 312
  worm, 311
Gear ratio, 309
Gravitational potential, 70
Gravity, 333
  definition, 68
  orbits, 81
    ellipse features, 82
    Halley's Comet orbit model, 85
    Inner Solar System model, 90
    Kepler laws, 83
  potential surface
    Algol model, 79
    Earth-Moon system model, 71
  printing tips
    Halley's Comet orbit model, 96
    Mercury and Earth orbit models, 95
  universal gravitational constant, 69
  vis viva equation, 68

H

Halley's Comet orbit model, 85
Hybridization, 251

I, J

Inner Solar System model, 90
Interference patterns, 52
Interferometry, 55

K

Kepler's Laws, 83
Kutta Condition, 105

L

Lift coefficient, 106
Light waves, 39, 332

M

Molecules, 335
    carbon atom model, 253
        assembling, 256
        printing, 246
        quantum numbers, 245
    crystals, 263
        diamond, 269
        water ice, 263
    electrons, 238
    hybridization, 251
        sp hybridization, 252
        sp^2 hybridization, 252
        sp^3 hybridization, 252
    noble gases, 242
    orbital shapes, 242
    periodic table of the
        elements, 239
    printing tips, 271
    valence electrons, 241
    water molecules, 259
        carbon *vs.* water molecule
            model, 262
        water molecule model, 259

N

National Advisory Committee on
    Aeronautics (NACA), 112
    angle of attack, 111
    camber line, 117
    dihedral features, 112
    four-digit profiles, 114
    swept wings, 109
    tapered wings, 109
    thickness equation, 118
    in World War II
        airplanes, 112

O

Octet rule, 241
OpenSCAD, 1
    download, 3–4
    editing, 4
    ideosyncracies, 13
    NACA four-digit profiles, 117
    plants (*see* Plants)
    simple machines, 155
    surface creation, 7
        Blocky one-sided surface, 8
        from external data file, 22

INDEX

OpenSCAD (*cont.*)
    flat-bottomed, 14
    3D printing, 16
    two-sided smoothed surface, 20
    trigonometric functions, 40

P, Q

Parker Spiral model, 63
Partial differential equations, 39
Plane waves, 43
Plants, 334
    botany, 196
        Camellia japonica, 199
        nutrients, 199
        sunlight, 198
        water, 197
    mathematics
        Fibonacci sequence, 203
        golden angle, 203
        golden ratio, 202
        meristem, 201
        phyllotaxis, 204
    OpenSCAD models, 204
        desert plants, 206
        flowers, 212
        jungle plant leaf models, 227
        plant/flower models, 217
        tropical jungle plants, 208
    printing tips, 230
Principia, 69

Principle of superposition, 43
Print-in-place, 314
PrusaSlicer, 24

R

Rack and pinion, 311

S

Simple machines
    compound machines, 155
    definition, 192
    friction and flexing, 155
    inclined planes, 156
        mechanical advantage, 153
        vs. wedge, 157
    lever
        class 1 lever, 161
        class 2 and 3 levers, 162
        variables, 166
    mechanical advantage, 153
    in OpenSCAD model, 155
    printing tips, 191
    pulley, 183
        assembly, 184
        mechanical advantage, 183
        variables, 184
    screw, 169
    wedge, 157
        frictional force, 157
        vs. inclined planes, 157

wheel and axle, 175
  assembly, 184
  variables, 184
Space weather, 63
Sprockets
  on a bicycle, 312
Surface creation, OpenSCAD
  Blocky one-sided surface, 8
  flat-bottomed, 14
    limitations and alternatives, 20
    printing considerations, 33
    saddle point structure, 8
    3D printing, 16
  from external data file, 22
  two-sided smoothed
    surface, 20

T, U

Torque, 305, 306, 309, 328
Trigonometric functions, 40
  coordinate system and
    conventions, 60
  diffraction, 49
    one-slit intensity
      function, 49
    Young's double-slit
      experiment, 52
  point sources and plane
    waves, 43
  principle of superposition, 43

printing considerations, 59
two interacting point
  sources, 47
Trusses, 336
  definition, 278
  models
    icosahedron, 298
    printing, 289
    tensegrity structure
      model, 289
    3-Rod tensegrity prism, 292
    2D truss, 283
  planar, 277
  tensegrity structures, 282
  3D truss, 281
  triangular structures, 279
  2D truss, 281

V

Vis viva equation, 68, 84

W, X

Water waves, 39
WWII Supermarine Spitfire
  model, 146

Y, Z

Young's double-slit experiment, 52